CITY ON FIRE

History of the Urban Environment
Martin V. Melosi and Joel A. Tarr, Editors

CITY ON FIRE

Technology, Social Change, and the Hazards of
Progress in Mexico City, 1860–1910

Anna Rose Alexander

University of Pittsburgh Press

Published by the University of Pittsburgh Press, Pittsburgh, Pa., 15260
Copyright © 2016, University of Pittsburgh Press
All rights reserved
Manufactured in the United States of America
Printed on acid-free paper
10 9 8 7 6 5 4 3 2 1

Library of Congress Cataloging-in-Publication Data

Names: Alexander, Anna Rose.
Title: *City on Fire: Technology, Social Change, and the Hazards of Progress in Mexico City, 1860–1910* / Anna Rose Alexander.
Description: Pittsburgh, Pa.: University of Pittsburgh Press, 2016. | Series: History of the Urban Environment
Identifiers: LCCN 2016007472 | ISBN 9780822964186 (paperback : acid-free paper)
Subjects: LCSH: Mexico City (Mexico)—Social conditions—19th century. | Mexico City (Mexico)—Social conditions—20th century. | Fires—Social aspects—Mexico—Mexico City—History. | Fire prevention—Mexico—Mexico City—History. | Social change—Mexico—Mexico City—History. | Urban ecology—Mexico—Mexico City—History. | Technology—Social aspects—Mexico—Mexico City—History. | Science—Social aspects—Mexico—Mexico City—History. | Social medicine—Mexico—Mexico City—History. | Mexico City (Mexico)—Economic conditions. | BISAC: HISTORY / Latin America / Mexico.
Classification: LCC HN120.M45 A44 2016 | DDC 306.0972/53—dc23
LC record available at http://lccn.loc.gov/2016007472

For Boris and Dotty Alexander

CONTENTS

ACKNOWLEDGMENTS

I became interested in fire history in 2008 when I was a graduate exchange student at the Benemérita Universidad Autónoma de Puebla. It was in Rosalva Loreto López's graduate seminar on urban environmental history where I first learned of this important subfield of history and the enormous analytical possibilities it has. While there I pursued a research project on fires in Puebla and realized that fire provides an unconventional angle to discuss larger processes in Mexican history. When I returned to Tucson, Bill Beezley, Bert Barickman, Martha Few, Katherine Morrissey, and Doug Weiner each encouraged me to examine fire from different vantage points, which made working on my project challenging, yet surprisingly fun. I am indebted to all of these mentors for their guidance and support throughout this whole process.

Uncovering the history of fire in archival sources was not always an easy task, and I have many people to thank for their thoughts on places to look and analytical avenues to explore. First, I would like to thank my colleagues and mentors at the Oaxaca Summer Institute, who offered suggestions about my project when it was in its earliest stages. There, Gabriela Soto-Laveaga and Claudia Agostoni motivated me to branch into the fields of public health and medical history. While I was in Mexico City, Ricardo Mendez Cantarell at the Archivo Histórico del Distrito Federal and Miguel Ángel at the Archivo Histórico de la Facultad de Medicina were always eager to help me and often brought to my attention sources I would not have considered otherwise. With the help of a Hagley Library grant in Wilmington, Delaware, I was able to consult sources about fire insurance and fire technologies. Additionally, Brian Freeman was the one who suggested I check out the Patentes y Marcas collection at the Archivo General de la Nación. Without his suggestion, this project probably would not have a technological focus at all. In my final stages of completing this manuscript, Tracy Goode and Justin Castro were kind enough to secure permissions and take photographs of documents, while the Department of History at Georgia Southern University generously offered to pay for the image reproductions for this book.

I am grateful to the many people in my life who have spent time discussing the historical significance of fire with me. Thanks to Lance Ingwerson, Tyler Ralston, Kira Robison, Cory Schott, Jessica Stites, and Steve Zdatny for influencing the way I think about Mexico, fire, and so much more. Several colleagues were kind enough to read over drafts. Thank you to Finbarr Curtis, Lizeth Pérez, Matthew Vitzk, and Andrew Konove for your comments and insights. Steve Bunker at the University of Alabama, Christoph Rosenmueller at Middle Tennessee State University, and Tony Rosenthal at the University of Kansas invited me to their campuses to present parts of this research. In each case, the comments I received helped immensely as I revised the manuscript. I would also like to thank the series editors, Joel Tarr and Marty Melosi, and the acquisitions editor, Sandy Crooms, for their help in shaping the manuscript. Moreover, I appreciate the attention to detail and historiographical insights of the two anonymous University of Pittsburgh Press reviewers who helped advance this book to the next level.

I owe an enormous debt of gratitude to my undergraduate advisor, Steve Lewis, at California State University, Chico. After one day of sitting in his course on Mexican history, I was hooked on Latin America and have never looked back. Finally, my family has been incredibly supportive. Through this time-consuming process, my family was always there to lend a hand or give me space when I have needed it. Thanks, Mom and Dad, for your understanding, and for visiting me while I was in Mexico City. Most importantly, thank you to my thoughtful and generous husband, Ryan, who has read too many drafts of this book.

CITY ON FIRE

INTRODUCTION
MODERNITY AND ITS ACCIDENTS

In Mexico until now fires have not been so terrible; the fine construction of our buildings does not allow fire to spread far; nevertheless, these disasters have been occurring more frequently, to the point that the precautions being taken are useless.

"BOLETÍN DEL 'MONITOR,'" *EL MONITOR REPUBLICANO*, SEPTEMBER 28, 1882

The simple combination of heat, oxygen, and fuel brings fire into living form. Its flames, the visible manifestation of fire, are capable of mesmerizing beauty and incalculable destruction. From the moment of its birth, fire fights for survival. The necessity of staying alive requires that it breathe oxygen and consume materials. At times it rages out of control, quickly transforming itself into a menacing force, taking drastic measures to fuel its own existence. It darts around corners and surges through corridors, whipping tongues of flames onto physical structures and stalking unsuspecting bystanders, crackling and hissing all the while as it feasts on wood, cloth, and anything else it might claim for subsistence. Clamoring for survival, fire leaves ruin in its wake. Charred remains and debris document its life.

Humans often have a decisive role in determining fire's survival as well. When it has the potential for human benefit—to light streets, heat homes, or propel machinery, for example—people stoke it, tend to it, feed it. But fire is often not so easily contained. For all they have tried to master fire, humans learned, often through grim experience, that fire possesses a life force all its own and does not always behave as expected. Over time fire's role in human development has evolved, but along the way it has also become more destructive. This basic fact has compelled people to devise various ways to control it and harness it for power.

When humans began to feed fire coal and natural gases—fuels that once rested deep in the earth beneath layers of rock and sand—their experience with fire also began to change. Feeding on these fuels, fire burns longer and hotter. Enclosed inside chambers and machines for

1

industrial purposes, it becomes more susceptible to explosion. This transformation in the use of fire, as it took hold in cities across the world, began to alter human-fire relationships, as people were forced to understand and respond to it in new ways. On the one hand, fire, once it had been harnessed in applications like steam and combustion engines, contributed to the benefits offered by new and more efficient industries. On the other hand, those who used flammable and combustible fuels soon came to realize that the fire that had been fed with this type of food exhibited an exceedingly unpredictable and dangerous nature. By reacting to it, regulating it, writing about it, and mitigating it, people acknowledged its decisive influence in the decisions they were making about the environment in which they lived. It had the power to excite and frighten, to transform landscapes and lives, to expose and exacerbate social inequalities, and to bring ideas about safety and risk to the forefront of policymaking. Coaxing fire to benefit humankind morphed it into something more robust and forceful than ever before, ultimately altering the course of human history.[1]

In Mexico City, fire portended the dangerous consequences of modernity and urbanization. Fire, and subsequent reactions to it, altered the development of Mexico's capital during the transformative half century from 1860 to 1910. During this period, the city's residents felt the effects of unprecedented population growth and the development of more and more industry. Consequently, efforts to modernize Mexico City and fold it into the global industrial economy had the unintended consequence of elevating the risk of fire hazards.[2] Because Mexico City functioned as both the main laboratory for the country and the site where global and local knowledge met, this book examines the development of modernity there.[3] While fires had afflicted the capital for centuries, never before had the citizenry confronted such a drastic increase in the frequency and intensity of this hazard. The numerous workshops that opened throughout Mexico City from the mid-nineteenth century onward required extensive fuel and fire energy tamed inside machines, making fires more likely to erupt. Soap manufacturers used large amounts of animal fats, tanneries needed thick oils for tanning hides, and ice workshops utilized chemicals stored under pressure. Combustible fuels (with flash points above 100 degrees Fahrenheit), such as varnishes, kerosene, fertilizers, and dyes, became common accessories in daily life and could easily be found in open-air market stalls, in home kitchens, or on corner-store shelves. Flammable fuels (with flash points below 100 degrees Fahrenheit), such as gasoline and turpentine, were only beginning to become popular among capital residents during this period. These energy sources, intended to make life easier and to increase business profits, also made living in Mexico City far riskier.

Environmental historian Stephen Pyne refers to the changing nature of fire in this period as the nineteenth-century industrial fire regime, since a number of cities throughout the world were plagued by comparable problems brought on by similar economic and technological changes.[4] While Mexico City in this period exhibited many of the characteristics Pyne identifies as part of the industrial fire regime, such as rapid urbanization and industrialization, the fuels used on a daily basis differed slightly in Mexico City. Unlike Western Europe and the United States, which had transitioned from biomass energy to coal during the early eighteenth and early nineteenth centuries, respectively, Mexico made the energy transition to coal much later, around the 1870s.[5] According to a recent study on household fuels, the majority of Mexico City residents and workshop owners used coal sparingly, and instead used wood, charcoal, and electric power for daily use.[6] The transition to petroleum and natural gas did not occur in Mexico until the 1940s. Despite the difference in fuels, Mexico City's population growth and density, as well as its increased output and manufacturing, make Pyne's concept of an industrial fire regime a compelling framework for this study.

The rise in fire hazards created a collective sense of fear in the city. The dangerous side of fire, the side that incinerated homes and took lives, forced residents to adjust the ways that they lived their daily lives, conducted business, and behaved in the city. In the face of growing fire risks, ordinary residents did not sit idly by and watch fires wreak havoc. Rather, they actively shaped fire control and prevention. The experiences of interaction and debate over the issue of fire among residents ultimately affected both the spatial layout and the political and social dynamics of Mexico City. Faced with the daily risk of fire, residents from diverse backgrounds made the city accommodate their needs. At times this meant becoming involved in municipal politics; at others it meant creating businesses to profit from urban risk, or pioneering new medical procedures to deal with the increase in the number of burn victims. Through these and countless other measures, political officials, fire inspectors, firefighters, municipal engineers, lay inventors, professional physicians, and ordinary citizens collectively transformed their city in response to a new and unfamiliar environmental threat.

Examining Mexico's capital from 1860 to 1910 offers a new approach to understanding how the city's history unfolded. Rather than employing the political periodization of the Porfiriato (the period of oligarchic rule under strongman Porfirio Díaz from 1876 to 1911), which perpetuates the conventional wisdom that modernization and urban development in Mexico City represented the will of the dictator, this project instead asserts that fundamental infrastructural and public service developments took root earlier in the city's history, beginning with the Liberal victory at the end of the Reform War (1857–1861) and expanding through the French Intervention (1862–1867), Restored Republic (1867–1876), and the Porfiriato. Without discounting the importance of political centralization during this period, this book maintains that a diverse group of actors, rather than just the political elite, shaped the city at this pivotal moment. This study thus uses an unconventional point of entry, fire, to examine the major changes in economic development, scientific understanding, and technological innovation occurring both in Mexico and throughout the world in the late nineteenth century.

Even though accidental business and home fires continued well into the twentieth century, this project's chronology ends around 1910, with the onset of another change in fire regime. During the 1910 Mexican Revolution, which ousted Díaz and spurred ten years of sustained internecine warfare, fires increasingly occurred as byproducts of combat or as intentional acts of arson by rebel forces. Revolutionaries set fire to haciendas (large rural estates that symbolized oppression to many in the countryside) and municipal buildings and burned property records and debt documents. Arson had been somewhat common and burdensome for city development in the late nineteenth and early twentieth centuries, but for the most part arsonists hid behind the common occurrence of accidental fires to gain insurance money or to seek revenge on a foe. The revolutionaries' more overt use of fire changed it from an accident of modernity to a tool of revolution.

This urban environmental history argues that technical expertise arose to address the various aspects of life that fire had affected, leading city officials, engineers, physicians, inventors, theater workers, vendors, and insurance agents to reevaluate how they understood and interacted with the dangers and limitations of their physical world. In the United States, urban environmental history has become a well-developed subfield, and discussions of urban parks, industrial pollution, natural disasters, water shortages, and sewage disposal mark just some of the contributions to it. But in Latin America, historians have just begun to view the built environment through the lens of environmental history.[7] This book is about how fire hazards worked at multiple levels in urban society, motivating citizens in everyday life to expand the fields of science, engineering, medicine, and business in order to confront urban risk. While borrowing from global scientific trends, Mexicans also utilized local expertise and experience to produce knowledge that they used to confront fires. In Latin America's growing literature on the history of science, scholars have noted the importance of social conditions and everyday interactions in the production and transmission of knowledge.[8] Relegating knowledge to the laboratory or university discounts the important findings made by the people who had to experiment with ways to protect their lives and livelihoods from an increasingly hazardous environment.

This book analyzes fire as an active agent in much the same way that historians of medicine and health have argued that epidemic diseases have shaped the course of history and geopolitics.[9] Rather than evaluating fire as simply a passive element that has been shaped by human action, this book also takes into consideration how fire interacted with nonhuman agents, such as fuels, winds, building materials, and chemicals.[10] Urban fires were fused with both natural and social forces that combined to present imminent danger to Mexico City. Functioning as natural occurrences that are constantly shaped by human initiative, fires can alter landscapes by destroying the built environment and straining natural resources. From a social perspective, the fundamental changes spurred by increased fires often intensified already severe inequalities, as access to fire safety was distributed unevenly along existing lines of privilege. The interconnections between space and inequality meant that not everyone had the same opportunities to be productive, healthy, and safe citizens.[11]

Patterns of scientific and technological change have shaped urban modernization in Mexico City, especially in the period between 1870 and 1910. Historian Mauricio Tenorio-Trillo refers to Mexico's capital during this period as the "ciudad científica" (scientific city) to explain how residents applied science and technology to the city to solve the problems of daily life.[12] Until recently, most analysis of technology in Mexico has focused on political and economic history with a heavy emphasis on mining extraction, textile manufacturing, and railroad development.[13] In the past five years, however, scholars have directed their analysis toward the social construction of technology and, more specifically, the cultural significance of new technologies for creating modern societies.[14] This book draws on the emerging field of scholarship on the history of technology in Mexico by employing the methodology and sources of cultural history and using them to make broader claims about how technology influenced the course of urban life in a moment of significant structural change.

Mexico City in the second half of the nineteenth century pulsed with energy from its bustling economy and vibrant street life. In the half century from approximately 1860 to 1910, the city's population more than doubled (from about 190,000 to 417,000).[15] This population boom was the result of people flocking to the city from rural areas and abroad to find employment; by 1900, half of Mexico City's population had been born outside of the capital.[16] Foreign companies followed suit and infiltrated the capital looking for new, untapped markets. The city changed from being a quiet capital to a thriving metropolis. For some, the transformation represented a renaissance, but for others, it led to horrid living and working conditions. The combination of rapid and diverse urban population growth, new risks, and the limited access to new technology reinforced and even widened the inequalities of a city already known for great social and economic disparities.

An infrastructure of flammability developed around the midcentury, when a higher concentration of manufacturing appeared in the capital. Beginning around 1860, the levels of production of metal extraction and manufacturing began to recover after the intermittent warfare that had affected the country during most of the first half of the nineteenth century. In part, the Liberal constitution of 1857, which promoted the proliferation of manufacturing and business, made economic recovery a possibility.[17] By nationalizing ecclesiastical property, President Benito Juárez set in motion a property grab by entrepreneurs who bought up church lands and buildings and in their places established businesses, factories, and workshops in the capital.[18] Mexico City, rather than the historic manufacturing cities of Puebla and Querétaro, became the country's primary manufacturing hub. By creating a favorable political and economic climate for businesses to thrive, officials hoped to concentrate factories in the capital, where more businessmen could benefit from centralized public works such as hydroelectric and communication infrastructure and thus lower production costs. The majority of manufacturing in Mexico City was centered on small-scale production (soaps, oils, pottery, and spinning wool) that primarily satisfied the needs of locals. Historian Gustavo Garza estimates that in

1879, 91.3 percent of businesses in the capital were of the small-scale production type, and the other 8.7 percent represented forms of export-oriented industrial production that included paper mills, tobacco workshops, and textile mills.[19] Most production in the capital relied on energy sources from manual labor, hydraulic power, and steam power, the latter requiring extensive use of wood from the foothills of the basin of the valley of Mexico. Because Mexico had so few coal reserves, coal never fully replaced biomass as it did in England and the United States. Instead, the coal Mexicans used merely complemented the existing fuel sources.[20]

With more manufacturing, a dense population, and new fuels in the city, fire increasingly became dangerous and exacerbated preexisting social divisions. This led to a series of political struggles, and official responses to disasters reflected a modernizing impulse that disregarded the material condition of the poor. Conservatives, feeling threatened by the Liberal reforms that limited the power of the church and military, tried to restore their political and social presence by first starting a civil conflict that would come to be known as the Reform War, and later by requesting help from Napoleon III of France. The emperor responded with the French Intervention in 1862 and in 1864 provided Mexico with a puppet emperor, Maximilian von Habsburg of Austria. Maximilian and his wife, Carlota, arrived as emperor and empress on the false assurance that Mexicans had consented to their presence through a plebiscite. With this in mind they sought to fulfill their duties to their new subjects, which included an effort to improve living conditions and beautifying the capital, even at the expense of depleting the treasury. Maximilian installed gas lamps in the center of the city, planted trees along avenues and boulevards, reinitiated garbage collection, and prohibited residents from dumping urine and human waste from balconies.[21] The emperor modeled many of his improvement efforts in the capital after similar projects in France, addressing social welfare issues with a vigor that had not been seen in Mexico ever before. Shortly after he arrived, he toured Mexico City schools, jails, and hospitals, discovering that they were understaffed and poorly maintained.[22] In addition to the emperor's observations, travelers and residents complained of foul odors and overcrowding in the city's hospitals and cemeteries.[23] Shocked by the insurmountable levels of poverty in the city, Maximilian and Carlota used their personal funds to expand public welfare, hospitals, and poorhouses.[24] The imperial couple's social welfare programs angered the conservatives who had brought them to power because they resembled earlier efforts made by the Liberal government. No matter who held political power in Mexico—Liberals, Conservatives, or an Austrian archduke—the city had become a dangerous place.

Fire risks became most visible in 1866, the year in which an exceptional number of conflagrations plagued the capital. Every few weeks, Maximilian's government received news of yet another devastating fire. Butcher shops, match factories, bakeries, and soap manufacturers all succumbed to fire during this remarkably devastating year. Sometimes the causes of the fires were clear: unsupervised lit candles falling over and setting the room ablaze,[25] for instance, or a poorly constructed oven setting aflame the walls of a bakery.[26] At other times the fires had no explanation at all: a box of matches spontaneously igniting in a drawer of clothing,[27] or an underground supply of gas exploding for no apparent reason.[28] In the existing scholarship on fires, and natural disasters in general, these small, daily occurrences have not received the same attention as the so-called great fires, such as the Great Chicago Fire of 1871 or the San Francisco earthquake fire of 1906. In the Latin American scholarship on disasters, numerous studies have used a major environmental catastrophe to reveal social tensions and racial and class-based dynamics that had been bubbling below the surface.[29] Conversely, in this project, it is the daily presence of fire, both big and small, that tells stories about struggles for power, safety, and resources.

The fear of fire in daily life did not emerge only in the aftermath of major conflagrations. Rather, it built up over time when communities faced smaller, more frequent, everyday fires. Instead of examining one major disaster, this book analyzes how the almost constant presence of smaller fires acted as a catalyst for social change. This analysis borrows from a branch of hazard studies that focuses on common disasters, even those that often were not counted in official records and are thus difficult to quantify. Greg Bankoff has led this approach with his work on the Philippines, which faces nearly constant threats of flooding, typhoons, mudslides, and earthquakes. The presence of these hazards in the lives of South Pacific Islanders led him to coin the term "cultures of disaster" to explain the ways in which certain societies live with constant natural threats and find ways to cope in the aftermath of misfortune.[30] Building on the idea of a culture of disaster, this book argues that urban residents, in a moment of significant social and environmental change, adjusted their daily lives to confront the increasing risk of fires. The small, everyday fire plagued cities across the world. One

Chicago underwriter warned of "the seriousness of our 'ordinary' or 'small' fires," citing in 1910 "an average of one conflagration a day" in the city.[31] Scanning Mexico City fire reports and newspapers reveals a similar statistic there. Preparing for ordinary fires required vigilance and regulation.

Once Porfirio Díaz came to power in 1876, the ministers representing his regime expanded fire suppression programs at the federal and municipal levels, but the reforms tended to benefit the upper ranks of society.[32] This unequal distribution of resources and support is characteristic of the style of liberalism that typified the Porfirian regime.[33] Sometimes referred to as "conservative-liberalism," the paradoxical Porfirian-style liberalism stemmed from the intellectual trends that influenced Mexican elites at the time, including Comtean positivism, Spencerian social Darwinism, and classical liberalism. This patchwork of theories provided the intellectual base with which officials adopted policies to distribute the benefits and risks of new urban developments across society.[34] In 1877, shortly after Díaz took power, all oversight of charitable organizations came under the control of the Board of Public Welfare (Junta Directiva de Beneficencia Pública).[35] Maintaining that the poor should not be coddled and therefore should not receive public assistance, officials refused to fund many existing programs that offered charity to the poor.[36] Officials chose, instead, to invest public service funds in very visible, large-scale projects, such as drainage systems and tree-lined streets, or in other conspicuous relocation and gentrification projects to achieve similar aesthetic and modernizing goals. The Díaz administration, in order to clean up the capital and rid it of poverty, moved lower-class hospitals and cemeteries to the edge of the city. These projects represented efforts to beautify the city and physically segregate the rich in the city center from the poor residents on its outskirts.

The Porfirian regime's belief that aligning Mexico with European sensibilities and aesthetics, especially the orderly and functional centers of Paris and London, would allow Mexico to achieve its evolutionary potential. This sentiment influenced the way the Díaz administration approached urban issues. Intending to make Mexico City the "Paris of the Americas," Díaz's advisers adopted urban aesthetic designs from the French *Beauté* and US-based *City Beautiful* movements.[37] The enthusiasm for European pastimes, attitudes, and technologies has been called the "Porfirian Persuasion," a phrase that refers to how upper-class Mexicans eagerly accepted fashion styles and sporting events from the United States and Europe and rejected already established popular domestic diversions such as bullfighting. They justified this new approach as a way to put the country on par with the seemingly progressive Western societies of Europe and North America.[38] The imitation of European aesthetics solidified a growing trend that embraced urban life, equating urbanization with modernity. From this new attitude emerged an urban ethic that inspired wealthier Mexicans and foreigners to move to the capital, where they invested in business and reveled in the city's cultural attractions. Historian Katherine Bliss describes Mexico City during the age of Díaz as the "playground of the Porfirian elite," a place where they could be seen eating truffles and bonbons from French chocolatiers and window shopping along avenues that housed boutiques full of the latest Parisian fashions.[39] The opportunity to attend the opera or sip coffee in upscale cafés attracted many outsiders to move to the capital.

Efforts at modernizing the city and making it appear more like its European and American counterparts contributed to the increase in fire hazards. Manufacturing workshops used large supplies of combustible fuels such as varnishes, sulfur, and turpentine, which could cause substantial destruction if ignited. The street lighting, praised by locals and visitors alike, was extremely dangerous if flames touched the stockpiles of gas or turpentine used to illuminate lanterns or if sparks flew from electrical apparatuses. Electric energy was introduced to Mexico in 1879, and over the next decades small plants began appearing throughout the capital, slowly replacing steam power with electric power. By 1890 all of the gas and turpentine-powered lamps in the city had been replaced with electric lights, and by the turn of the century more than half of the electricity that powered the country was both generated and used in Mexico City.[40] Newly established parks and forests, intended to bring health to the population, often caught fire in dry months or during lightning storms. In addition, building patterns, especially experiments with wooden construction, mimicked those found in Europe. Whereas planners formerly used flame-resistant cobblestone and tiles, they increasingly installed flammable wooden sidewalks to line the streets and constructed French-inspired mansard roofs that adorned the mansions and hotels of upper-class neighborhoods. Mansard roofs were particularly prone to fires. Joseph Bird, a Boston fire expert, warned in 1873 that "mansard-roof structures, as made in our cities and villages, are the most dangerous buildings ever constructed . . . they will assuredly cause the destruction of our cities."[41]

Mexico City planners did not heed the warnings made by Bird and others and instead continued to build the steep wooden roofs that now symbolize Porfirian Mexico City.

Modernization had its price, and Mexico City inhabitants looked at urban beautification and improvement projects with varying degrees of pride and concern. Petitions to the Ayuntamiento (municipal government), newspaper articles, and legislation confirm that residents of the city worried about the precipitous increase in fires.[42] Everyone from vendors to bureaucrats speculated about the reasons for this drastic increase. Some turned to religion to understand these dangers and ultimately justified their misfortune with the explanation that God had meted out punishment for sins committed.[43] Others blamed population increases for the increase in fires, often embellishing their arguments with class-based slurs about the effects that uncultured rural dwellers had on the city.[44] Stories of recent migrants to the city who lived in one-room homes crammed with seven or eight people, dogs, chickens, pigeons, and pigs alongside piles of charcoal and wood, exemplified concerns that rural inhabitants were incapable of adjusting their daily habits to fit the modern urban environment.[45] Moreover, the modernizing context characterized by land dispossession, job opportunities in the city, population density, and increased manufacturing contributed to the creation of a new industrial fire regime.

This project utilizes an assortment of primary sources to bring breadth and depth to the study of fire. Reading the opinions of reporters, government officials, artists, and travel writers reveals how residents and visitors alike understood Mexico City to be a hazardous place. The rationale behind fears of fire was rooted in both the presence of real incidents of fire and the imagined fears that the city was a tinderbox. In either case, there was growing sentiment that the city needed order and control. In the process of combing through these sources, the voices of capital residents who were most affected by fire emerged. While I had initially anticipated the centrality to the story of certain social actors such as fire engineers and firemen, other voices took me by surprise and led me to archives I never would have entered otherwise. Inventors and physicians, for instance, represent some of these valuable, yet unexpected, individuals. Patent requests, inventors' drawings, medical journals, and medical school curricula confirm that fire was not only a concern of the political elite. The responsibility of preventing and suppressing fires and healing those who had been burned preoccupied citizens from multiple backgrounds, and this newfound concern about fire safety created new occupations and ways to earn a living. Because fire affected so many disparate facets of everyday life, this book as a whole is organized thematically, rather than on a strictly chronological basis. The narrative structure moves from large-scale, often intangible perceptions and fears of fire and ultimately ends with a discussion of flame and smoke's effects on human cells and tissue. This approach helps to focus the glance, starting from the macro and ending with the micro, thus explicating the importance of fire to residents on various levels.

Popular depictions of fire as unruly, menacing, and evil stoked popular imagination and helped create a collective fear of fire. Stories of fires in Mexico and abroad swayed public opinion and encouraged civic engagement. Chapter one focuses on fear, both of fire and how people responded to it, ultimately confirming that fear functioned as an engine of change. The chapter argues that residents used emotionally charged pleas to inspire change and make public officials give priority to urban safety. With the considerable increase in fires, traditional informal and community-based approaches to fighting fire, such as the bucket brigade, could no longer contain the bigger and more frequent fires. Instead, citywide ordinances and a professional fire brigade became necessary. Chapter two analyzes a changing ethos among citizens about the government's responsibility to regulate behaviors in the city. Both urban planners and public health officials used the science of regulation to try to bring order to disorderly spaces in the city, even going so far as to reach into people's private homes and classify some daily habits as risky. Fire codes defined fire hazards as detrimental to business, public health, hygiene, and safety, and the codes eventually divided the city into zones of comparatively great and comparatively mild fire risk. However, ordering spaces and regulating behavior was not enough, and city officials had to implement fire control practices in case the preventative measures did not work. Chapter three discusses the establishment of the vital social service of a professional fire brigade. These uniformed men with the newest imported technologies from Western Europe became emblematic of an orderly, progressive city.

A professional fire brigade was just one of the fire-related occupations that arose in the last decades of the nineteenth century. Chapter four examines how university-trained engineers and architects, functioning as city inspectors, implemented the fire codes that government officials had authorized. Engineers used their technical expertise to manipulate their natural surroundings to prevent fires. This culminated in the creation of extensive hydraulic systems to supply water for drinking, industry, and fighting fires. But over time, the presence of engineers also created a false sense of security by convincing people that the city had become impervious to the risks of fire. Thus, paradoxically, people actually became more reckless (refusing to follow official fire codes, for instance), despite clear evidence of the increased risk and frequency of fires. Engineers' experiences as inspectors made them recognize that fires threatened all members of society, and they argued that fire protection should be extended more equitably. Their appeals for fair distribution of protection marked a major deviation from the Porfirian mindset, which tended to value elite progress at the expense of the poor.

During a period when residents turned to science and technology to improve social problems, and when the spirit of entrepreneurialism was increasingly celebrated, lay inventors listened to the growing anxieties about urban danger and created safety devices for homes and businesses. As chapter five discusses, they vigorously marketed their protective services—safety matches, flame-retardant roofs, and handheld fire extinguishers—to make a profit from fire risks. The presence of Mexican inventors challenges the erroneous notion that in Latin America technological innovations were always imported from the United States or Western Europe. Yet inventors of technologies were not the only actors to profit from fire risk. Chapter six evaluates the ways in which insurance representatives, playing on the fears of fire, promised security against potential loss of investments. They saw fire as an opportunity to sell peace of mind in the face of catastrophe. Insurance companies reinforced fears of uncertainty, loss, and death by telling clients that their lives were inherently vulnerable and could suffer complete demise at any moment. By purchasing insurance, Mexico City residents attempted to predict and prepare for emergencies, refusing to leave anything to chance. Moreover, private insurance fit into the liberal economic model that privileged free market capitalism and encouraged an ethos of individual responsibility. Businessmen or homeowners who could afford private insurance were more willing to take investment risks because they had a safety net.

Moving beyond fire's effect on the built environment or the economy, chapter seven assesses how fire hazards caused pain and suffering to the human body. Burns became easily infected, and more severely burned patients rarely survived more than a few days. For those patients who survived, fire marked their bodies with unsightly scars and the trauma of the event haunted their memories. Physicians experimented with a combination of indigenous healing methods and laboratory-developed medicines, as well as skin grafts from animals and cadavers. The increasing numbers of burned patients forced physicians and healers to make healing burns part of their professional mission.

During the period from 1860 to 1910, fire and fire safety marked the city in irreversible ways. Fire, an anthropogenic agent that can destroy structures and incite fear, changed human–nature relationships in the growing metropolis of Mexico City. Fire hazards offer a way to look at broader processes found in rapidly modernizing cities. They demonstrate how space is made and remade according to political and social agendas, how public services and technology get distributed unequally, and how the competing economic and political interests of private and public interest groups are reconciled with the collective necessity to create a safe environment. In other words, fire forced different groups, through varying measures of conflict and cooperation, to grapple with their hazardous environment and assert their interests in discussions about how best to confront it.

CHAPTER ONE
FIGHTING FIRE, FIGHTING FEAR

Fire, as a servant and friend, is useful and agreeable; as a master and an enemy, fire is a tyrant and a destroyer.

HENRY L. CHAMPLIN, *The American Firemen*

Stories of a gruesome fire in February 1866 circulated in the Mexico City press, provoking a collective sense of dread that lingered for several months. The fireworks shop on San Antonio Abad Street, which held a large supply of explosive powder, had suddenly and without explanation burst into flames.[1] The blaze incinerated five members of a family of firework makers and left the wife of the owner with her flesh burned to the subcutaneous layer, exposing bone and muscle on her chest and face. Bystanders rushed her to the Hospital de San Pablo, where they treated the deep and extensive burns that left her mangled head hairless and her body blackened by charred skin. No one expected her to live more than a day or two, but she nonetheless hung on for three days after the conflagration, enduring excruciating pain. Maximilian, the European emperor then governing Mexico, and his wife, Carlota, were so astonished by the tragic news that they provided the few surviving widows and orphans with several hundred pesos to rebuild their lives and pay for the burials of their husbands and fathers.[2] This tragedy affected the imperial couple to such a degree that they reversed their previous insistence that street cleaners use fire pumps for watering the plants in Alameda Park and cleaning the Ayuntamiento buildings. Henceforth, the pumps were to be reserved for emergency situations only. The episode demonstrates the extent to which fire had become a political concern, and officials ranging from low-level bureaucrats to the emperor offered suggestions about how to avoid such a massive disaster in the future.

Material changes to the built environment increased fire risk in the city and affected how citizens responded to hazards. Prior to the 1860s, fire protection, like various other social services in the city, had been a neighborhood responsibility. When a fire erupted, the first person on the scene ran to the nearest church to ring the bell, signaling that neighbors should rush to help suffocate the flames. While the neighborhood firefighting system worked well enough for centuries, with the introduction of highly flammable and often explosive agents residents frequently made comments like this one about a fire in March 1866 that erupted in a shop that sold gas: "The efforts to stifle it were useless, because from the beginning the fire was very lively."[3] The increasing presence of "very lively" fires fueled by gases, gunpowder, or fireworks meant that residents needed more than volunteer efforts and buckets of water to extinguish the flames. The concerns of uneasy businessmen, mothers, shop owners, and foreign investors motivated officials to manage the changing urban environment more effectively.

A culture of fear developed in response to fire hazards in the capital. Expressions of popular culture, ranging from newspaper articles to broadsides to ballads, circulated widely and evoked a sense that the city was a dangerous place. At times, the sense of fear assumed paranoid dimensions and residents questioned all aspects of fire control and prevention. Could neighbors be trusted to practice fire safety privately within their homes? Did city officials have a plan to handle major disasters and the manpower to implement such a plan? Were some areas of the city safer than others? This growing fear of fire motivated residents to intensify their pleas for a professional firefighting brigade. A typical strategy in letters to the municipal government or newspaper articles included detailing the terrifying sensory experience associated with fire hazards. Personal testimonies of the sounds of a hundred voices yelling "fire," the piercing screeches from policemen's whistles, or the clamor of church bells heralding fire all brought life and emotion into appeals for municipal support.[4] Others detailed the unusual beauty of flames enveloping buildings, describing how flames in the distance looked like a sunset made of violent rays of crimson and purple.[5] Engrained into their memories, these experiences haunted people long after the fire. Months and years later people could recall in precise detail the smell of the burning building or the sensation of heat on their skin.[6] Whether or not they had experienced a fire firsthand, residents often recalled stories about fire in their communities, and those tales fed collective anxieties and ultimately inspired officials to act.

Stories of conflagrations at home and abroad affected how some people thought about their future. Imagining flames engulfing one's home or business often brought anxious feelings about lost investments or, even worse, excruciating death. Newspaper descriptions of burning victims jumping out of third-story windows to escape flames or somber reports of stiff, asphyxiated bodies found in the aftermath of fires heightened uneasiness about urban fires.[7] Certainly, individual fears based on personal experience or psychology existed, but collective experience and memory helped to determine which daily activities became interpreted as risks.[8] The fear of fire found its way into popular culture and popular politics alike. Printed and oral stories, vividly recounting fires that took place throughout the world, piqued the imaginations of entire communities. Residents of Mexico City pieced together information about great fires, scorched bodies, and lost homes, and transposed their fears of such horrors onto the history of the city. Citizens, paralyzed by fear, imagined the city as an unsafe, disorderly place full of vice and danger.[9] Whether it was fear of crime or fear of fire, the collective angst that developed throughout the city motivated citizens to use emotionally charged pleas as tools to get officials to initiate public fire protection.

Newspaper articles, broadsides, and ballads contributed to growing anxieties about urban living. Days after the fatal Paris Opera fire of 1887, which took the lives of more than two hundred people, the Mexico City–based newspaper *El Monitor Republicano* published a lengthy article questioning the safety of the capital's theaters. The article used the Paris experience as a way to launch a larger debate about fire safety in Mexico City.[10] Similarly, for several days after the 1906 San Francisco earthquake and fire, the front page of *El Imparcial* was covered in articles that depicted the disaster, retold personal tragedies, and described the relief measures implemented in a city consumed by flames and buried under rubble. The coverage of the catastrophe in San Francisco even included articles about how Mexican citizens, many of whom had family and friends living in California, had been affected.[11] One Mexican account of the San Francisco earthquake and fire described it in the following way: "What dreadful confusion! What indescribable screaming! Some were asking for help and praying to be saved, and all were overcome by the greatest panic one could imagine."[12] Reporting on the horrors of big and small fires around the world became a standard practice in most major Mexican newspapers. Stories of the rapid destruction of more than one thousand homes in Shanghai in 1894, or of the sounds of screaming patients locked in their cells during a fire at a mental asylum in Montreal, entered Mexican homes and made citizens question their own fire safety.[13]

While Mexico City never had a conflagration on the scale of the Great Chicago Fire of 1871 or the Paris Opera fire of 1887, the frightening and descriptive reports reminded Mexico City residents that fire could affect anyone.[14] Many fires during this period could be characterized as accidental—a neighbor leaving a candle lit or electrical lighting sparking in a match workshop—and thus difficult to predict or contain. Nevertheless, they often quickly spread to adjoining structures, making the threat a concern to entire communities.[15] One example involves Alejandro Hernández, the twelve-year-old son of the owner of a fireworks shop, who caused great panic among his neighbors when a spark flew from one of his fireworks onto a pile of lumber and quickly incinerated the business. As the workshop burned, pieces of pyrotechnics shot out from the windows and doors in every direction while neighbors looked on in horror and prayed that they could put out the fire before it spread to adjacent buildings.[16] Such horrifying tales made it seem as though anyone in society could be the victim of someone else's oversight or unwise decisions.[17]

On at least one occasion the fear of fire transmuted into a full-scale urban legend, showing how panic could become distorted by rumor. Rumors spread quickly about a phantom killer known for dousing unsuspecting victims with flammable oils and lighting them on fire. Whispered stories about a shady Jack the Ripper–type arsonist who prowled the streets of Mexico City with an oil can and a box of matches eventually made their way into the written record (newspapers and minutes from Ayuntamiento meetings) and became an issue for government officials. A police investigation confirmed that on two separate occasions, two women had burned to death after their dresses caught fire, but they assured the Ayuntamiento that this was not something that should concern capital residents. In an attempt to quell fears and squash the unfounded rumors, Mayor José María Icaza gave a series of public speeches and ordered municipal officials to hang explanatory posters throughout the capital reprimanding those individuals who had spread such "miserable speculation" and generated "almost universal terror" among the population.[18] As with most urban legends, no one knows for sure if such a character ever existed. No matter the source of the rumor or how true it might have been, these exaggerated stories had the power to circulate quickly and instill fear in the citizenry. But while rumors certainly had the power to generate fear, it is also important to note that rumors often came into being as a product of fear. Rumors about fire spread quickly because at the time it was not inconceivable to imagine burning to death. Therefore, rumor was both generative and reflective of a culture of fear that had developed in the capital.

For those convinced that God would punish a sinful population with floods, plagues, and other natural disasters, the fear of fire was even more arresting. In an article entitled "The True Causes of the Earthquake on November 2, 1894," the author explains, "it is incredible that Divine Providence punishes all his creatures for the criminal behavior of some of his children," and goes on to describe the insurmountable crime and vice that had been eroding morality in Mexico City.[19] A similar religious explanation for fire came from a young man who had been badly burned after a chemical reaction in a latrine created an explosion. The physician who treated the burns mocked his patient for believing that the flames that scorched his body came from the gates of hell opening beneath his feet.[20] Others also made jokes about superstitious explanations for disasters. In a newspaper article about the San Francisco earthquake fire of 1906, a Mexican journalist in jest questioned whether God had punished San Franciscans with the tragedy. He ultimately pointed out the absurdity of such spiritual logic and instead blamed the earthquake on the fault line that sits below the city. The article ended with an ominous warning that the natural world contained unforeseeable risks that threatened all humanity. For many residents, prayer was one way to deal with tragedy.[21] Broadsheet images by artist José Guadalupe Posada often depicted scenes of fires, floods, and earthquakes and confirm that many people relied on religion and prayer as their primary tools to confront natural calamity. In one of Posada's broadsides about an 1894 earthquake, he captured the turmoil and pandemonium of a city plaza being shaken by a temblor. In the background of the image the buildings are askew and lightning bolts dart through the sky, while in the foreground the townspeople are portrayed kneeling in prayer or asking for forgiveness by extending their arms open to heaven.[22]

Prayer and religious offerings traditionally offered some citizens a way to cope with or explain disasters. Since the seventeenth century, Mexicans have documented their personal faith through small devotional images, typically painted onto a piece of tin, known as votive offerings (specifically, *retablos*, *ex-votos*, or *milagros*). These modest thanksgiving offerings include narrated pictures of saints performing miracles. In this expression of popular Catholicism, grateful people thank a saint for intervening in their lives during a time of turmoil, often after falling ill or becoming injured in an accident.[23] Votive offerings have become a valuable resource to historians who are interested in studying daily life in Mexico, because the images document everyday suffering, ailments, and accidents and allow viewers to peek into cultural spaces that are usually inaccessible in other types of sources.[24] Surviving an accident—whether being struck by lightning, falling off a horse, or catching fire—is a common theme in these offerings. For example, after a fire erupted in Don Silverio Aguilar's kitchen, his family feared that the flames would reach the thatch roof and spread to the rest of the house. Yet, through what seemed like a miracle, the fire stayed in the kitchen and the family was able to extinguish the flames with ease. The family attributed this miracle to the Lord of Sacromonte and commissioned the painting of a votive offering to tell the story.[25] Similarly, in 1902, gunpowder manufacturer Cirilo Ramirez commissioned a votive offering to thank Señor de la Buena Muerte for healing his three children who had been badly burned during a house fire.[26] Religious explanations for fire suppression have persisted and remain common today, but as fire expertise developed and the use of firefighting technologies became more widespread, some residents gave thanks to both divine intervention and technological innovation for surviving a fire.

While some suspected that horrific fires could be explained as God's wrath, this fatalistic interpretation waned as intellectual trends rooted in liberalism, positivism, and social Darwinism became more widely accepted. Slowly, municipal authorities, writers, physicians, and other formally educated groups came to embrace an ideal based on the conviction that reason and science could explain calamity and make sense of the unthinkable.[27] By the mid-nineteenth century, with access to new methods in seismology and technologies to combat disasters, discussions of the causes of natural disasters entered the realm of the scientific and moved away from the religious.[28] Yet these new urban technologies elicited new and powerful fears of their own. During the heyday of Porfirian development, residents witnessed impressive but unsettling technological, scientific, infrastructural, and architectural developments enter the city. These new and modern developments often increased urban risks. While some residents eagerly awaited new technologies, others expressed concerns that they would bring more fires to the city.

Some residents found new lighting technologies disquieting. Through much of the nineteenth century, gas lamps served as the main sources of lighting throughout the city. To provide fuel to light the lanterns, engineers had to place underground gas tanks below houses, hospitals, and theaters. At times this modern amenity came at a grave cost. In March 1866, one of these underground holding tanks ignited and burned down a corner store and several houses, killing a night guardsman and injuring two others in the process.[29] Beyond the risk of fire, gas lighting had other troubling attributes. A report about hygiene and lighting complained that gas lighting "smelled bad, was hard to see in, heated up the air, consumed a lot of oxygen, could occasionally lead to asphyxiation, and caused explosions and fires."[30] Electrical lighting marked a major improvement over gas. Advertisements from Mexicana de Luz y Fuerza, a domestic, private electricity corporation, tried to convince capital residents to purchase electric lighting for their homes and businesses by arguing that electricity obviated the need for matches. Never again, the ads suggested, would consumers need to clean candle wax or ash off furniture. Most importantly, the company explained that electricity carried no risk of fire, a blatantly erroneous claim in light of the frequent fires that originated in light bulbs and sockets.[31] For example, at a celebration for the Virgin of Guadalupe in 1895, festival organizers adorned the church with Mexican flags, clusters of flowers, and ornate cloth, and illuminated the altar with electric light bulbs. Clergy chose to abandon the traditional candle flame, often associated with life and vitality, for the safety of electricity. Once the crowd had packed into the church and commented on the splendid lighting display, a spark flew from one of the light bulbs and set fire to the hanging flags.[32] In a like manner, on at least two occasions department store window displays, which used electrical lighting to illuminate the products and lure in customers, shorted out and contributed to the nearly complete destruction of the La Valenciana (1900) and the Palacio de Hierro (1914). Modern technology such as electric lighting continued to ignite fires and alarm residents.

Railroads and streetcars, the modern marvels that made possible the rapid transportation of goods and passengers across vast distances, also caused deadly fires. Nearly once a week newspapers reported train accidents and fires, which helped intensify fears of fire and fears of technology.[33] Few Mexican trains had been fitted with spark arrestors to prevent embers and sparks from flying off into the air. This meant that trains started fires at an alarming rate, often near towns and areas with fuel sources to keep the fire alive.[34] Sparks flying off the wheels often hit cotton or wood that had been stored at the station

Figure 1.1. José Guadalupe Posada, "Quemazón en el Baratillo de Tepito," 1913. Courtesy of the Jean Charlot Collection, University of Hawaii at Manoa Library.

ready for shipment.[35] When a train passed through Puebla in March 1890, a spark flew off one of the wheels and hit the home of Manuel Concepción César, completely consuming his property.[36] In 1911 a train full of people experienced a grisly fire when a tank of petroleum suddenly exploded, splattering the blood of several dead passengers and dozens of chickens all over the train station deck.[37] At one point the chief of the firefighting brigade refused to allow the construction of a new tramway that ran on a street occupied by a number of valuable homes and businesses. He knew of the streetcar's propensity to ignite fires and did not want expensive buildings to catch fire—yet he had no objections to the tramway running through lower-class neighborhoods.[38]

The most severe criticism of technology appeared in the broadsheet images by José Guadalupe Posada. Posada, known for his satirical illustrations of political life, depicted appalling scenes of trolley cars, buses, and trains falling over or plowing into innocent bystanders. His portraits of daily life not only exposed his personal criticisms of technology, they also reflected the social anxieties of ordinary Mexicans.[39] These drawings often accompanied newspaper articles and *corridos* (popular storytelling ballads) and helped to spread widespread anxieties about the ill effects of new technology in a rapidly modernizing society.[40] His rendering of a fire in the lower-class neighborhood of

¡¡LA GRAN DESTRUCCION
Y TERRIBLE INCENDIO
DE LA PLAZA DE TOROS DE PUEBLA
EL 12 DE ENERO DEL PRESENTE AÑO!!!
¡¡¡Un Muerto, Muchos Heridos y Contusos!!

Figure 1.2. José Guadalupe Posada, "¡¡¡La Gran Destrucción y Terrible Incendio de la Plaza de Toros de Puebla. El 12 de Enero del Presente Año!!! ¡¡¡Un Muerto, Muchos Heridos y Contusos!!" in *Posada's Popular Mexican Prints*, xiii.

Tepito (one that did, in fact, occur) shows faceless men dressed in white linen either running out of a smoke-filled building or lying motionless as flames envelop their limp bodies. Accompanying the horrifying image of destruction and certain death was a corrido that recounts how the blaze ignited most of the products in the market, leaving many vendors destitute. Reports filed after the fire noted that all sixteen market stands had been reduced to ashes.[41]

Folklore and ballads bemoaned nature's conquest over manmade structures while at the same time providing seamy accounts of human willingness to profit from tragedy or put someone else in danger. Before Tepito residents could completely extinguish the fire in the market, neighbors, onlookers, and firemen looted valuable items that had not been engulfed in the flames.[42] The uproar caused by fire hazards did much more than destroy physical structures; it also, according to popular lore, inspired thievery and other, horrific crime. A corrido about the 1902 fire in Puebla's *plaza de toros* (bullring) detailed a nightmarish scene of the human capacity for selfishness and cruelty. As soon as the fire erupted in the packed stadium, the spectators screamed and ran around frantically looking for an exit. As they pushed and shoved their way through the entrance to escape from the heat and smoke, some onlookers resorted to bludgeoning other audience members with chairs and stones in order to reach the exit unharmed. The author of the corrido, disgusted by the scene, described the participants as hellish monsters and demons, claiming that the audience members of Spanish descent had caused most of the destruction.[43] Cramped spaces tended to escalate risk, spurring residents to label theaters, bullrings, and other venues that held spectators of multiclass backgrounds as unsafe spaces. Printed and oral stories depicting terrifying scenes of flames devouring people's flesh or mobs trampling over innocent victims reiterated to city dwellers that they had much more to fear than just the flames, given how vile, cruel, and selfish their neighbors could be.

Uncovering emotion in archival data can reveal how fear and grief affected the daily practices of the living, prompting people to take action or ascribe ideas about safety and danger onto space.[44] Petitions to Mexico City's Ayuntamiento about fire hazards contain the most conspicuous and poignant expressions of fear in the archive. Personal testimonies about fires, descriptions of neighbors hoarding dangerous combustibles, and complaints about unimplemented fire laws all contain hints of anxieties about the future. After a fire consumed his neighbor's home, one concerned citizen wrote to the city council to alert officials of the likelihood that another fire would occur on his street because many of his neighbors stored flammable and explosive materials in their residences.[45] Another citizen, Juan de Zárate, addressed the Ayuntamiento about his unease in the aftermath of a large fire at a phosphorous plant, claiming that the fire ignited because the owner of the manufacturing workshop did not follow the 1829 fire code. Zárate then went on to cite numerous other cases of business owners in the city who refused to obey the law and would likely bring more death and destruction to the city.[46] More and more residents began to follow Zárate's lead and wrote the municipal government to make known their fears that neighbors would unintentionally start fires. For example, one resident who lived above a bakery and had to deal with constant heat rising up to his apartment eventually requested that the city engineer inspect the business downstairs to ensure that no fire threat existed.[47] Similar to the concern about the bakery, the majority of the complaints contained accusations of neighbors' negligence and disregard for public safety and looked for some kind of authority or expert to address the issue.

The last half of the nineteenth century marked a period in which virtually all parts of the world underwent urbanization and experienced increases in fires. Mexico City was no exception. Its attempts to imitate the building styles, electric lighting, and transportation systems of Western Europe and the United States ultimately made the capital more prone to fires. Through frightening and emotional experiences with fire, people prepared for disasters and looked for security. Rather than simply await death or debilitation, they effectively pleaded with the municipal government to offer some protective options. By voicing concerns about unrelenting fire, everyday citizens dramatized the formidable danger of some spaces while praising the inherent protection that others offered.[48] Most importantly, since as one official explained it, "nothing is more frightening in a fire than disorder," capital residents wanted to bring order and control to the city as a way to prevent terrible fires from occurring in the first place or, conversely, to manage them effectively if they happened to erupt.[49] To bring order to the city required the assistance of authorities who could regulate harmful behaviors.

CHAPTER TWO
SCIENCE OF REGULATION

An ounce of prevention is better than a pound of cure.

JOSEPH BIRD, *PROTECTION AGAINST FIRE*

Mexico City's transformation into the "ciudad científica" began in the second half of the nineteenth century when, as historian Mauricio Tenorio-Trillo explains it, "the city and science acquired a scandalous coexistence." He claims that the scientific management of Mexico City was evident in "the massive sanitary transformations, city enlargements, and replanning of large capital cities,"[1] but this way of looking at how science shaped the city gives primacy to large, visible infrastructural transformations. By looking at science through the lens of fire hazards, it becomes clear that there were also much smaller, scientific tweaks or adjustments to the city and its people that contributed to the making of the ciudad científica. In response to citizens' concerns about fire, which arose out of both a real increase in the number of fires and popular culture that amplified fire-related anxieties, officials used the science of regulation to pacify concerns.

By focusing mainly on fire prevention techniques, officials established fire zoning laws that placed the onus of fire safety on residents, which led officials to regulate and penalize the behaviors of capital residents. Public health officials, immersed in the growing literature about hygiene, applied the logic of sanitation and disinfection to fire hazards. These experts tried to decontaminate the city by applying preventative solutions that would reduce the number of fire-related illnesses and ailments. They framed industrial hazards as something that could be controlled, anticipated, and prevented.[2] This framing mechanism coupled fire control with social control, giving public health officials grounds to monitor and regulate behavior all in the name of health. In all instances of fire regulation, the regulatory efforts tended to focus on the center of the city, where the most expensive homes and businesses were located. Therefore, the spatial focus of this legislation reflected and in turn reinforced Mexico City's unequal class structure.

FIRE ZONING LAWS

A comprehensive set of fire laws had been in place in Mexico since 1777, with the passage of a royal decree for preventing fires. These late-colonial laws, which emanated from Crown authorities in Spain, outlined architectural requirements for ovens and roofs, restricted the locations of fireworks shops and wood storage facilities to the suburbs, and fined residents between ten and fifty pesos for violations.[3] In 1829, just eight years after Mexico achieved independence, the municipal government of Mexico City reviewed this colonial legislation, adding several new articles and increasing the fines for violations.[4] Despite the existence of fire codes, residents complained that officials did not enforce the regulations. For much of the nineteenth century, government officials had bigger concerns—violent internal political disputes and equally disastrous foreign invasions—leaving municipal authorities with little manpower and few resources to regulate fires. Patrolling public buildings to ensure that they contained working water pumps or fining people who stored excessive amounts of wood in the center of the city were little more than afterthoughts in an era of political and economic instability. Nevertheless, as fires started to consume more buildings and as the nation became more stable and the state more capable, officials regarded fire regulations as fundamental to maintaining order in the city.

As early as 1860, there had been some discussion about installing a professional firefighting brigade in the capital, but by 1862, when Mexico was engaged in a war with French troops, officials understood that such an undertaking would have to wait. A fire department required expensive equipment, along with hundreds of full-time employees who would require training, all of which municipal officials could not secure. The solution that the Ayuntamiento proposed would cost only a fraction of the cost of creating a fire brigade; this solution was simply to enforce the existing fire codes and draft new fire-related laws. José María González Mendoza, a brigadier general and governor of the Federal District, developed numerous regulations that assigned specific duties to citizens, based on their occupation, in the instance of a fire breaking out. This decree directly targeted water carriers (aguadores), policemen, night guardsmen, and physicians, and gave general good-Samaritan guidelines to anyone who happened upon a fire.

González Mendoza's regulations attempted to codify how people were to react to fire danger. The first witnesses to the fire were required to shut any doors or windows to the building in order to prevent breezes from spreading the flames to nearby buildings. After that initial step, the law required that at least one witness run immediately to the nearest church and ring the bell to alert policemen and night guardsmen. The night guardsmen, upon hearing the sound of the church bell, needed to quickly gather ladders, buckets, and any additional supplies that could help put out the fire and rush to the scene. The law also demanded that the city's water carriers fill their jugs and any other large containers with water to help extinguish the flames. While the water carriers and night guardsmen put out the fire, policemen were to control the crowds of people and make the onlookers whisper or speak softly so that the volunteers could hear when the chief of police shouted orders. Additionally, physicians and surgeons needed to be present to tend to any burned or injured victims.[5]

In theory, this form of fire management would work well because each resident had a specific task to undertake and this would eliminate confusion or dawdling at the scene of a fire. González Mendoza claimed that this method solved the city's problem of a lack of a firefighting brigade because it made assisting in fire suppression a mandatory part of urban life. To make collective firefighting strategies more effective, the law outlined a series of punishments and rewards. If someone neglected to fulfill his specific task or impeded the volunteers' efforts to smother the flames, the Ayuntamiento would fine him for disregarding public safety. On the other hand, if someone assisted in the humanitarian service of putting out a building fire or saving the lives of people or animals without giving in to the temptation to loot or rob the owner of the burning building, he would have his name printed in newspapers and would receive a medal of thanks or a small monetary reward for heroic efforts.[6]

González Mendoza's fire guidelines that designated specific tasks to specific people had mixed results. The availability of water, the proximity to church bells to call for help, and the location of the police department or the night guardsmen's posts all factored into the success of this plan. The devastating fire at the Mercado del Volador in 1870 exemplified what could go wrong with González Mendoza's regulations. When the fire erupted, the night guardsmen were asleep, the wells near the market had no water, the pumps held at the police department did not work, and the crowds of curious people interfered with volunteer ef-

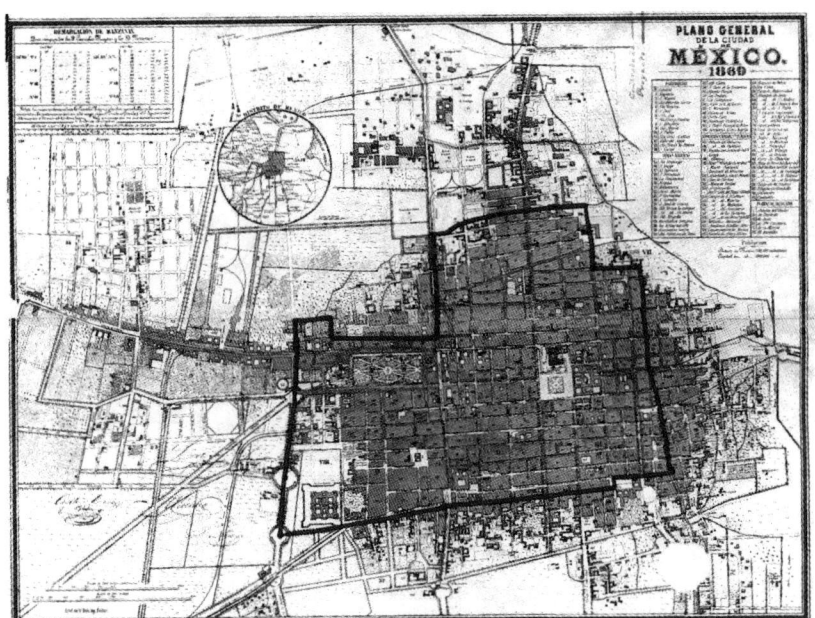

FIGURE 2.1. This demarcation shows the fire zoning law of 1871. Plano General de la Ciudad de México, Planero Horizontal 2, Gaveta 1, Plano 2, AHDF, 1869. For the fire code regulations that stipulated the center of the city and the suburban parts, see Art. 21, Los suburbios de la ciudad, September 26, 1871, AHDF, caja 41, exp. 10, fs. 5.

forts to put out the flames.[7] After the disaster, Ayuntamiento members began to rethink their strategy and questioned whether everyday citizens could actually help suppress fires, or if they only made fires worse.

Shortly after the fire at the Mercado del Volador, city officials adopted a law that sought to tackle fire hazards more efficiently by prohibiting the presence of combustible and flammable fuels in the city center. This fire zoning law, which delineated the center and periphery of the city, transformed the spatial configuration of the capital in remarkable ways. In regards to zoning, at times the officials mapped the boundaries of the city center based on the presence of drainage ditches or waterways, while at other times the urban-suburban divide ran directly through neighborhoods or along large avenues. Among other things, the center of the city included the cathedral, the Alameda, the Mercado del Volador, and many principal churches and theaters. This demarcation excluded Chapultepec Castle, farms, and lower-class, largely indigenous neighborhoods (figure 2.1).[8]

This expansive fire code made strict distinctions between the center of the city and the suburbs, in effect transferring major fire risks to the outskirts of the city in order to increase safety within the urban core.[9] After the fire code was adopted, residents living and working in the city's center were granted two months to replace any existing wooden structures with brick, stone, or adobe. If residents neglected to do so, they would be fined between 5 and 200 pesos or put in jail for up to ten days.[10] In an attempt to prevent further destruction, the law prohibited the construction of buildings taller than seventy-two feet and required fireproof materials to be used in construction.[11] Residents living in the city center also needed to remove stockpiles of wood and store them in suburban neighborhoods. Firework shops, which were especially prone to horrific and unrelenting fires, could only be located in the suburbs. Even *pulquerías*, working-class taverns that served the fermented agave beverage known as *pulque*, were prone to fire hazards and therefore required extra precautions that limited the amount of pulque that owners could have at the bar.[12] With the implementation of this legislation, residents constructed buildings differently, moved shops and industries to suburban areas, or changed their business practices in order to avoid hefty fines or prison sentences for disobeying the fire code. Making such drastic changes did not happen overnight, or for that matter even within the designated two-month period that officials allocated for residents to change their flammable habits.

Ayuntamiento officials claimed that they did not have the funds to hire and train firemen and instead chose the pragmatic and money-saving solution of regulating people's activities in the capital. Some citizens—carpenters, fireworks manufacturers, and matchstick makers, for example—suffered more than others due to the relocation of their businesses to areas with less social importance, away from prominent churches or wealthy neighborhoods. By framing safety as something that could be controlled and managed through appropriate behaviors, officials and reporters often blamed victims for not following the fire codes. When the adobe home of Juan Pérez caught fire, newspapers reported that Pérez's neglect made him lose the few humble pieces of furniture that he owned.[13] In a 1909 fire at a bookbinding shop, firemen cited carelessness as the reason that the owner lost most of his property to fire. Had he remembered to remove a pot of glue from the stove, they argued, he could have avoided the fire and not endangered his entire neighborhood.[14] In spite of the regulatory efforts aimed at preventing fires, these hazards nonetheless persisted, in part because residents found it onerous to change their behaviors or invest in expensive fireproofing materials. Other city officials tried a different approach to convince residents to abide by fire codes: they tied fire safety to health and hygiene.

FIRE DANGERS AS HEALTH CONCERNS

In response to a growing incidence of industrial accidents, officials in late-nineteenth-century Mexico City reconsidered traditional definitions of health. They decided that it should expand to encompass safety concerns more broadly and include hazards, accidents, and disasters as integral aspects of health and wellness. As a result, fire prevention efforts became enmeshed in a larger package of health initiatives that officials designed to make the capital safer and the population more productive.[15] Through the institutionalization of health programs, Pasteurian germ theory began to concern both physicians and bureaucrats alike. As they tried to control populations and redefine behaviors, they made compliance with health codes part of the fabric of daily life.[16] Fire safety fit within a larger intellectual discourse about health and particularly found a place within the increasingly important discipline of hygiene.

In this period, the concept of hygiene was considerably more expansive than its more restricted meaning today of general cleanliness. Instead, it encompassed a host of social and cultural factors including immorality, poverty, and disease, which all became clumped together in hygiene initiatives and treated like viruses that could easily spread and thus needed to be contained.[17] Mexican scientists and physicians crafted various definitions of hygiene over the course of a half century. In 1857 one physician, Roberto Gayol, defined hygiene as the science of the relationships between humans and the world around them and, moreover, elaborated that it represented the study of how this relationship could be improved to protect the health and extend the lives of humans.[18] Two decades later, another physician asserted that the study of hygiene needed a more meaningful definition, something that included ideas of "philanthropy" or "love of humanity," since hygiene was the science of honorable men who would do anything to preserve the population's health.[19] By the turn of the century, Dr. Julio Delobel expanded ideas about hygiene and referred to it as equilibrium: good hygiene meant that one had, in addition to strong physical health, intellectual vigor and moral integrity. He went on to describe how sanitary codes protected not only people's health but also their economic investments.[20] In a similar vein, Dr. S. Morales Pereira also reflected that good private and public hygiene translated into financial prosperity for the individual and the nation.[21] Over a half century, hygiene transformed from a concern about healthy human systems (or the physical manifestation of health on the body) to one of healthy social and economic systems. Hygienists understood that fire hazards could have ill effects on both the human body and society as a whole.

The scientific trends of bacteriology and germ theory inspired sanitary experts to try to metaphorically disinfect the city of potential fire hazards. Their attempts to predict and prevent fire hazards by regulating fire risks offer a clear example of the intersections of urban policy and public hygiene in a growing, industrializing city.[22] Along with the increasing number of manufacturing industries in the city, an early form of industrial hygiene emerged.[23] New technologies, equipment, chemicals, and substances posed direct threats to worker safety and health. While some businessmen and industrialists argued that workplace injuries were a necessary part of living in a technologically advanced society, others recognized the highly dangerous nature of factory equipment as detrimental to quality of life.[24] Industrial accidents offered residents a way to derive meaning from the influx of technology into their daily existences, and the fear of occupational hazards represented just one of the ways that people understood the role of technology and industry in their lives.[25] This increasing social awareness of industrial hazards prompted officials, especially inspectors, to anticipate potential hazards, regulating everything from airborne particles to building construction to workers' behavior.

Concerned that people did not know how to care for themselves properly, the national and municipal governments wanted public health officials to become more present in the everyday lives of ordinary citizens. In discussions of the high mortality rates in the capital, health experts often blamed residents for uncritically continuing practices that experts believed caused harm to the body.[26] People feeding children milk from diseased cows, having unlicensed midwives deliver babies, living alongside human excrement and animals, or dressing children poorly in cold weather represented just some of the complaints made by elite members of society. The members of the Superior Health Council, established in 1841, acted as advisers to the federal government in health-related matters. The council hired only educated, well-respected men as health inspectors. This job required a degree or certification as an engineer or architect, at least eight years of service to the profession craft of engineering or architecture, and an impeccable reputation in official society. If a physician wanted to work for the council, he needed a government license verifying his educational training.[27] Roberto Gayol, an engineer of the National Board of Health, reflected on the responsibilities of sanitation inspectors and hygiene campaigns by explaining that this small group of men of science were the only ones who could answer the nation's most pressing social questions.[28] Due to the country's political and economic instability in the midcentury, the council had minimal authority until the 1870s.[29] Thereafter, especially under the Díaz regime, public health became increasingly important, and officials became convinced that having a healthy population meant that the country could have a vital and productive citizenry.

In 1882 the Superior Health Council experimented with public health campaigns at the municipal level before launching the National Sanitary Code of 1891, which became a paragon of sanitary reform across the Americas.[30] The Mexico City Sanitation Code of 1882 regulated all dangerous, unhealthy, cluttered, and crowded establishments in the capital and defined fire hazards as detrimental to public health, hygiene, and safety. The growing number of burn victims forced public health officials to build on existing regulations and tailor them to their health and hygiene objectives. Officials adopted the original demarcation of the city's center and periphery as defined in the 1871 fire code and adjusted it to take into account future growth, especially in the western part of the city where builders had started to construct the upper-class neighborhoods of Juárez, Cuauhtémoc, Roma, and Condesa. Public health officials coupled fears about foul odors and miasmas with existing fears about fires and explosions, thus making an even sharper divide between the city's core and periphery.[31]

The 1882 sanitary code, especially article 157, provided the most explicit directions for restructuring the city based on the calculation of fire risks, linking health with social change. The law categorized hundreds of industries into three groups based on the likelihood that they would cause fires. The first category listed industries that could operate only at a considerable distance from private houses or public roads. These included factories that produced oils, varnishes, fertilizers, and dynamite, as well as all slaughterhouses, petroleum refineries, and smelting operations. The second detailed all of the industries that had to be located in the suburbs, including steel mills, arsenic plants, adobe manufacturing, and match workshops. The third encompassed those industries that could be situated anywhere if they had a yearly fire inspection by the sanitation commission or the police department. Among these were breweries, bakeries, cigarette shops, coffee roasting houses, and soap plants.[32] The Superior Health Council redefined what spaces were healthy or unhealthy based on the types of activities that occurred in these establishments.

Foreign observers praised the national sanitation code, which brought Mexico international accolades for its hygiene initiatives. Its success was due largely to the efforts that municipal officials had taken during the previous two decades to refine and experiment with the regulations on a city level. Hygienists held regular health congresses in the capital, which allowed them to make adjustments to current health regulations before showcasing the country's sanitation code for an international audience.[33] In many ways, Mexico's municipal-level sanitation reforms acted as trial runs for the larger, nationwide sanitation code.[34] The 1892 American Public Health Association (APHA) conference, held in Mexico City, was an opportunity to display years of meticulously crafted health reforms to world-renowned physicians and hygienists. Upon returning from his trip to Mexico's capital, Dr. Irving Watson, the secretary of the APHA, published an article about Mexico's sanitation code, praising its comprehensive nature as the most advanced in the world.[35] The Superior Health Council was recognized as a global leader in hygiene and public health.[36]

To a greater degree than earlier fire codes had, the health department's regulations focused on the behaviors of workers in hazardous work environments. The law prohibited them from smoking cigarettes, lighting cigars, and even wearing shoes with iron nails, which could create sparks if struck against the ground with enough force. In workrooms and warehouses that held flammable substances, workers had to cover the workroom floor with a thick layer of sand, fill buckets of water each day for fire control, and dig ditches that were to be filled with sand around the outside of the building to provide a barrier in case a fire did erupt. These regulations also targeted the lighting in manufacturing workshops and prohibited the use of artificial lights in workshops with flammable materials unless they were safety lights. For workshops that operated at night, all lighting had to be placed outside, next to glass windows or doors.

Factory regulations and the updated 1891 sanitary code tried to make ordinary citizens into whistleblowers by encouraging residents to report any activities they deemed dangerous or unhealthy. Some citizens wrote to the editors of newspapers warning other members of society to stay away from dangerous parts of the city known for fires or explosions.[37] Other residents went directly to the Ayuntamiento or Superior Health Council to divulge information about the activities and practices of their neighbors and businessmen in the city. The sanitary code promised that if anyone made a complaint about a potential fire hazard, bothersome noise, foul smells, spilling liquids, or excessive smoke, the municipal government and the Council on Public Health would immediately suspend the business activities of that establishment.[38] At his earliest convenience, that zone's health inspector would arrive on the premise to inspect the veracity of the claim and demand that the owner make improvements. After receiving complaints about a match workshop called Señores Lascuráin y Compañía, the council shut down the operation due to a long list of violations of the health regulations, ranging from phosphorus poisoning to fire risks.[39] In a similar scenario, a workshop on Niños Perdidos Street that produced acids and other chemical substances was shut down temporarily after health inspectors deemed it unsafe because the building sat too close to other structures, the chimney for the oven did not have a wide enough opening to let toxic airborne particles escape, and the oven's head would likely catch the wooden roof on fire.[40] In addition, theater owners received double criticism from not only city engineers but health inspectors as well. The presence of fires in theaters could lead to air poisoning because fires in rooms with poor ventilation depleted the oxygen, suffocating people with carbon dioxide.[41]

The Council of Public Health made information about the safety in factories and businesses public, an attempt to adjust consumer practices and make customers consider whether or not they wanted to purchase goods from owners who had not followed official safety regulations. Violations of the sanitary code were published in the newspapers El Diario Oficial and El Municipio Libre, thus exposing the unsafe practices of owners who put their workers and the rest of their neighborhoods in danger.[42] Each business had to post its license in a visible place so that customers and inspectors could see it. Businesses also received public recognition from the health commission when they followed health ordinances. In 1889, the Council of Public Health visited a bakery called Señores Albeitero y Arranche. Pleasantly surprised by the excellent sanitation condition in the bakery, health inspectors gave a laudatory report to the council, which was eventually published in El Estudio and the Diario Oficial. Claiming that the Señores Albeitero y Arranche bakery should become a model of excellence for others to emulate, the inspectors explained that the business had abundant lighting, good ventilation, and clean restrooms. In their estimation, it exemplified the hygiene, safety, and cleanliness that the council sought to promote. To allow readers to comprehend just how exemplary this establishment was, the article described a parallel situation of a bakery that was poorly lit, wet, and dirty, with old machinery, insufficient ventilation, and a wooden floor that could easily catch fire.[43] The article thus offered a cautionary tale that encouraged consumers to buy products from companies that invested in safety.

A healthy population, according to public health authorities, had the potential to be a productive population.[44] In 1912 Dr. Winthrop Talbot tried to convince manufacturers and industrialists that sanitation, hygiene, and workplace safety helped generate profit.[45] In a speech to the APHA in Washington, DC, Talbot explained that "humane dealing becomes good business. Charity and philanthropy have no place in industry, but kindliness is not charity, nor is generosity alms-giving. . . . Sanitation is a means of saving dollars and cents." Rather than using a moral argument about the treatment of workers, Talbot attempted to persuade business owners to consider safety by framing it as a problem of cost. As the industrial hygiene movement picked up in the early twentieth century, employers had to start taking responsibility for occupational hazards by paying fines or compensating injured workers.[46] The double-edged sword of worker injury and government fines drove many employers to adopt the latest trends in factory safety.[47]

Health reforms benefited a variety of people. Government officials touted the country's sanitation reforms as a way to showcase its progress. Some business owners embraced the reforms because it made their workers more productive and decreased risks. Workers, housewives, businessmen, and other residents wanted to achieve long, happy, healthy lives and the hygiene campaigns offered the most concrete ways to prevent the spread of disease and the possibility of hazards. Rather than using an economic argument to promote health, Dr. Eduardo Licéaga, in the preamble to the 1891 sanitation code, framed healthcare as a humane act. Licéaga referred to citizens as brothers and claimed that protecting their physical condition and overall health and happiness also meant protecting the country.[48]

Fire prevention legislation represented one part of the growth of a competent bureaucracy that tried to protect the city and its residents from misfortune.[49] Bureaucrats from the Ayuntamiento and the Council on Public Health used legislation as their tool against fire, attempting to mitigate fire hazards by regulating the most dangerous and potentially hazardous establishments. They sought to separate safe and unsafe spaces in the city, always justifying their actions as rooted in the science of health and safety. The work of these public servants meddled in the lives of residents who did not necessarily want to change their daily behaviors, tear down wooden structures and build brick ones, or succumb to public ridicule if their businesses did not pass inspections. Preventative measures, while somewhat helpful, could not stop fire hazards in a moment of such rapid industrial development and urban growth. The continuing fire problem, coupled with the nuisance of fire regulations, forced residents to demand more expensive and visible methods for achieving urban safety, specifically the establishment of a fire brigade. Citizens' fears, memories, anxieties, and experiences intersected with the municipal government's objectives for a progressive, modernized, and safe city. In the face of residents' emotionally charged requests for a firefighting brigade and the accompanying technology to support the brigade, officials, no longer able to rely exclusively on preventative measures, were forced to embrace the newest approaches to hazards management.

CHAPTER THREE
CONTROLLING THE FLAMES— THE FIRE BRIGADE

The evil flames rise:
Fire! A hundred voices shout,
a shrill whistle and hoarse cries.
The terrifying church bell.
Startled and confused
the city is shaken to its core.
Ready! The pumps fly open,
the flames touch the heavens!
The fire grows, the smoke rises.
Sparks fall, embers rain;
there is nothing to be done.
Come on! More power! More water!
The debris burns and burns;
and among the smoke there stands
the silhouette of a fireman.
Radiant, imposing, fearless.

"EL BOMBERO," *EL MONITOR REPUBLICANO*, DECEMBER 9, 1883

During the first several decades of the nineteenth century, paid and trained firefighting brigades became fixtures in major metropolitan centers such as Paris, London, and Boston. This municipal service symbolized modern urban life and functioned as a necessity to protect the lives and businesses of city residents across the world.[1] For Mexico, establishing a firefighting brigade was an emblem of modernity and safety, a sign that trained experts were prepared to control seemingly uncontrollable fires. But by the first decade of the twentieth century, firefighters had gained a mixed reputation. While some residents were confident in the fire brigade's abilities, others were actually more discomfited by their presence. Many citizens felt firemen were corrupt, incompetent, and dangerous. From the perspective of officials, the development of a fire brigade was simultaneously a good-faith effort at municipal reform and an act of political theater.

APPEALS FOR A BRIGADE

Mexican newspapers reported on the heroic acts of firefighters around the world, often characterizing them as saviors in an era of unprecedented risk and danger.[2] One newspaper article claimed that all the civilized nations and best cities of the world had top-rate fire departments, but Mexico City had nothing to protect the budding industries and valuable buildings that had been constructed in the capital.[3] Several outside observers wrote letters to the Ayuntamiento in an effort to convince officials that the capital desperately needed a fire brigade. Francisco Schiaffino sent the city council a copy of the official firefighting guide for Paris, which included pictures of firemen in uniform, explanations of the function of pumps and other equipment, and details of the salaries and pensions that each member received.[4] Several years later, Juan Turín, an instructor of gymnastics in Guadalajara, offered his services to the capital. Turín explained that firemen in the United States and France received gymnastic training to improve their agility, stamina, and strength, and if the Ayuntamiento were to fund a gymnastic academy for firemen he would happily move to Mexico City and train the men.[5]

Word had spread about the successes of other cities' firemen, and nearly every postfire report in Mexico City ended with a notation that had there been trained firefighters at the scene, they would have reduced the damage. In response to these suggestions, city council members asserted that the municipal funds could never cover the costs of pensions, equipment, funerals, or hospital visits.[6] By the early 1850s the municipal budget could not afford to repair the fire pumps, which had been used to attack US soldiers during the war between the two countries in the previous decade, let alone fund an expensive initiative to develop a professional fire department.[7] Instead, the Ayuntamiento took steps to start a volunteer brigade that would require far fewer resources. The first real attempt to initiate a fire brigade occurred in 1856, when the governor of the Federal District, Juan J. Baz, attempted to recruit men to serve the city. The requirements called for men between eighteen and forty years old, who were strong, healthy enough not to become easily fatigued, and stood at least five feet tall.[8] This initial attempt proved unsuccessful because too few men volunteered. Officials soon realized that in order to get enough men to join the firefighting cause, the Ayuntamiento would have to allocate funds to create a professional brigade that offered more incentives to firemen, including salaries, pensions, and workman's compensation.

The French *sapeurs-pompiers* became the model for the Mexican fire brigade.[9] After studying the French system, in 1860 Francisco G. Casanova, the commander general of the Federal District under the Juárez government, drafted an ambitious set of bylaws for the creation of two professional firefighting brigades in the capital.[10] Similar to both the British and the French firefighting departments, the Mexican equivalent would operate under the control of the police department. The two brigades would comprise 110 men, most of them artisans, to be guided by captains who had been trained as architects or engineers. All men who wanted to become firemen had to demonstrate abundant physical vigor and stamina. To prove that they could be trusted, candidates had to provide character references that attested to their reliability and good behavior. Casanova and other officials knew that the dangers of firefighting would deter men from joining the brigade, so they included provisions for pensions and life insurance for men injured or killed while on the job. As late as 1900, a Russian delegate at the International Congress of Firefighting criticized the French *sapeurs-pompiers* model for being too expensive for most municipalities to afford, which was precisely the problem Mexico City faced in the 1860s.[11] Similar sentiments about the costly French model came from Cuban officials who chided governments in Europe and the United States for devoting large and "sometimes wasteful" sums of money to pay the salaries of professional fire brigades. Cuban officials alleged that men in Havana had natural inclinations toward public service and performed their "philanthropic and humanitarian duty with such devotion and enthusiasm" that Havana's volunteer fire brigade, which was established in 1835, easily rivaled the professional brigades found in New York, London, and Paris.[12] Even though other Latin American cities such as Santiago, Buenos Aires, and São Paulo had great success with their volunteer brigades, Mexico City officials came to the conclusion that a professional brigade on par with the French *sapeurs-pompiers* was the right model for the city.[13]

Drawing up legislation for a professional brigade did not mean a swift implementation of the law or an immediate reduction in the level of risk. A decade after Casanova drew up his legislation, a professional brigade still did not exist and fires continued to ravage the city. Instead, inexperienced auxiliary policemen, night guards, and water carriers took charge of fire suppression because there was no professional brigade and too few men wanted to be part of the volunteer brigade. Municipal officials complained openly that no one stepped up to save the city from destruction. As José María del Castillo Velasco, governor of the Federal District, noted in 1871, "In the United States and Europe, the most distinguished young men make up the fire departments. Perhaps in Mexico this can be achieved, because there is nothing more philanthropic and respectable than putting out a fire."[14] The New York Underwriters Agency confirmed this by stating that there were always more than a thousand applicants waiting to enter the New York City Fire Department.[15] In 1871, the same year the governor complained about how few Mexicans volunteered, he tried to persuade Mexico City residents to assist in fire suppression by offering nominal monetary compensation. After each fire, the city council would award ten pesos to anyone who had attempted to put it out or assist in saving lives.[16] While the Ayuntamiento criticized citizens for refusing to become volunteer firemen, citizens simultaneously criticized officials for not having done enough to develop a fully equipped, professional fire brigade. It took several decades to recruit men, determine pay scales, and purchase the requisite equipment, but a professionalized fire brigade was eventually realized in 1888, making Mexico City's one of the first professional firefighting departments in Latin America.[17]

PORFIRIAN-STYLE FIREFIGHTING

The shape that the brigade finally took reflected the broader vision of the Porfirian regime. In this period, Mexico City became the showpiece of national development and the laboratory of modernization projects, and shortly after Porfirio Díaz took office, his administration started to funnel money into municipal services as a way to promote the mission of order and progress.[18] Among other things, that mission courted foreign investment, offered protection to industrialists, and portrayed the country as a stable place to live and conduct business. The fire brigade bolstered that mission. As with most Porfirian developments, the fire brigade had an air of modernity and progressiveness but also helped to reinforce the severe inequalities of the period.

To bring order to the city, Ayuntamiento officials provided policemen, inspectors, night guards, and firemen with substantial funding to subdue chaos in the capital. Policemen, inspectors, hygienists, and firemen all had a similar goal: to prevent the eruption of chaos, whether it came in the form of violence, robberies, or disorder and confusion during a fire.[19] In his study of crime and criminality in Mexico City, Pablo Piccato explains that public services, especially the police force and the penal system, provided government officials with a way to extend their reach to the street level. The policemen or city inspectors who roamed the city streets monitored residents' behaviors and taught them to act appropriately in the capital.[20] Firemen, too, helped extend the control of government officials and the visibility of firemen reminded residents of the flammability of the city. To increase the visibility of public servants, officials supplied soldiers, policemen, and *rurales* (rural police forces) with uniforms and state-of-the-art weapons (Winchester rifles, cannons, and dynamite), which conveyed authority in society.[21] At the same time, government officials awarded firemen similar accouterments: European-style uniforms, imported equipment (fire engines, hoses, and pumps), and the right to enforce fire codes and collect fines.

The presence of a paid and trained brigade patrolling the streets with the latest pumps and engines reflected turn-of-the-century trends in urban improvement. As in other cities throughout the world, the constant presence of a uniformed and professional fire company was aimed to relieve citizens' fears about fire.[22] Mexico City followed this same trajectory: by making firemen a fixture of city life and equipping them with imported uniforms, helmets, ladders, and fire engines, the municipal government made it known to citizens that it cared about fire safety. Nevertheless, the buildup of a firefighting apparatus was slow. Even though municipal authorities had anticipated that the brigades would employ more than 160 men in 1878, almost three decades later, in 1905, the three fire stations staffed only seventy-seven men and a half-dozen officers to protect a population of roughly four hundred thousand people (one fireman for every five thousand residents). In comparison, by 1887, New York City had 1,300 firemen and officers in its fire department to protect a population of almost 1.5 million (approximately one fireman for every 1,100 residents),[23] and by 1910 Paris had a brigade of 1,800 firemen to protect its population of 2.7 million (approximately one fireman for every 1,500 residents).[24]

In Mexico City, the 160 firemen seemed paltry in comparison to the two thousand policemen employed to patrol the city streets by the end of the Porfiriato.[25] Nevertheless, their presence paid off on numerous occasions. In December 1883 a laudatory poem about Mexico City's firemen appeared in *El Monitor Republicano*. It described the hazy, debris-filled scene of a fire, but beyond the smoke stood the figure of the fireman—"radiant, imposing, and fearless."[26] Later, in March 1887, a pile of hay outside of La Industria, a ceramics studio, caught fire and spread flames throughout the workroom. Bystanders noted that only nine minutes after the alarm had sounded, a brigade of firemen arrived on the scene and suffocated the flames completely, saving the surrounding buildings.[27] On this and many more occasions, residents expressed an outpouring of gratitude for what the fire brigade had accomplished. The brigade and its accompanying technology represented a visible expression of security for the capital and ensured that foreign investors and businessmen felt confident in establishing industries and businesses there.

For the most part, public services tended to benefit the wealthiest sectors of society. The locations of the three fire stations within the city offer an example of the spatial disparities of urban safety. All of the fire stations were near Alameda Park, which left residents of eastern neighborhoods such as Tepito without quick fire control services. A similar trend can be found among policemen, who primarily patrolled the central district, an area frequented by wealthy patrons and foreign tourists. The placement of fire departments and policemen near the center of the city reflected a materialist relationship between class and property. In the aftermath of fire, the opulent government buildings and first-class theaters would cost more to rebuild, and therefore those properties seemed worth protecting. Yet the prevailing ideologies of social Darwinism and Comtean positivism may also help explain a cultural aversion to providing equal protection to the rich and the poor. The social Darwinist idea that the poor were destined to remain poor ultimately helped officials justify why they distributed public services unequally. When combined with Cometean positivism, which rested on the idea of predetermined stages through which each society must pass in order to achieve progress, the poorest populations would hold the rest of society back from achieving progress and modernity. By cutting the poorest classes off, Mexican officials made a material statement that they could bring the country to a position of global prominence more quickly.

Whether or not everyone received fire safety, most residents welcomed the fire engines and pumps as a comforting presence in a context of rapid urbanization and heightened risk. While many residents during this period took forms of technology, especially transportation technologies, as signs of danger that should be approached with caution, the technologies of fire safety elicited a different reaction. Instead, fire safety technologies were liberating. They were freedom from fear and freedom from dealing with environmental problems. However, they did not always work as well as promised. One afternoon in May 1861, the upper level of the post office caught fire, and even though there had been a portable pump on site and dozens of men to help put out the flames, the fire raged on into the night.[28] Companies from Europe and the United States accused Mexico City of not having the most up-to-date equipment and quickly took advantage of new attitudes toward hazards that encouraged the importation of fire control technologies.[29] By employing fear tactics in their advertising campaigns, these companies disseminated the idea that Mexicans remained unprepared to handle fire-related catastrophes. The representative for the Patterson Machine Company of New Jersey sent his company's brochure to the Ayuntamiento with a letter warning that Mexico City's outdated and inefficient fire pumps would never be able to put out any major fire.[30] Other fire equipment companies detailed the devastating fires around the world that could have been prevented with their supplies.[31] These tactics multiplied the already growing concerns about conflagrations while sowing the seed of belief that Mexico City residents had been deprived of the newest approaches to urban safety.

The municipal and national governments responded favorably to the demands for new firefighting technologies. Between 1873 and 1875, the country imported fire pumps from Germany, France, Great Britain, and the United States.[32] When in 1900 an electrical short caused the walls and roof of the department store La Valenciana to burn, the fire brigade employed the help of a fire engine, and they put up a valorous effort to stop the conflagration. Unfortunately, the building eventually crumbled and Sebastian Robert, the owner of La Valenciana, had to rebuild his store from the ground up.[33] Nevertheless, in a broadside accompanied by a corrido, the firemen and the fire engine received public recognition for their heroic duties (figure 3.1).[34] Mexico City officials kept purchasing these technologies. In 1905 when Sanborn Maps, the world's leading producer of fire insurance maps, conducted an extensive appraisal of fire safety in the capital, mapmakers reported that Mexico City's three fire stations collectively contained two Merryweather steam fire engines, two smaller Merryweather steamers, seven hand engines, 1,500 feet of waterproofed cotton hose, and six-

Figure 3.1. José Guadalupe Posada, "La Quemazón de la Valenciana," n.d. Courtesy of Yale University Art Gallery.

teen horses to pull all of the pumps and engines.[35] Not until 1914 did the fire department purchase its first gas-powered fire engine (figure 3.2). Despite these improvements, the number of firefighting supplies in Mexico City remained small in comparison with those found in major European and American cities.

Regardless of the lower than expected number of firemen, fire stations, hydrants, and pumps, these improvements to public safety dotted the city with signs that the municipal government was committed

FIGURE 3.2. Company of Mexico City firemen in their fire engine. They nicknamed the engine the "King of Fire," after Ramón Corral, the Vice President of Mexico who served under Díaz from 1904 to 1911. *Overland Monthly*, 117.

to protecting capital residents and capital properties. These efforts helped curb fire hazards but nonetheless contained elements of political theater to show residents that the government cared about safety concerns. In 1906, Félix Díaz, the nephew of Porfirio Díaz and inspector general of the police and fire departments, supervised a fire test in the center of the city to showcase the work of the fire brigade. After the fire brigade set a large makeshift wall on fire between Tacuba and Atzcapotzalco Streets, more than one thousand spectators watched the flames grow and the clouds of smoke billow in the plaza. Once the blaze had become sufficiently large and intimidating, a brigade of firemen unrolled the heavy hose and shot a torrent of water at the flaming wall.[36] This publicity stunt, intended to prove that the capital's fire brigade could skillfully extinguish unruly fires, ended with roaring cheers from the amused crowd. Two years later President Díaz and Vice-President Ramón Corral gained international fame for a similar publicity stunt, though in this case in response to a real danger. When a fire erupted in a municipal building that held important government documents, Díaz and Corral posed for pictures while they handled a fire hose pointed at the flames.[37] Newspaper articles reported on firefighting spectacles but also retold the sacrifices firemen willingly endured and ultimately reassured citizens that firemen were honorable public servants. After fireman Doroteo González survived falling from six stories from a ladder while trying to put out a fire, *El Imparcial* printed his picture and thanked him for his patriotic service.[38] While these were not entirely hollow gestures on the part of the municipal and national governments, officials created media hype about firefighting that ultimately connected officials to safety initiatives.

Showcasing safety extended beyond newspaper articles and firefighting demonstrations. The uniformed firemen, alongside policemen, military officials, and rurales, held prominent positions in public parades and celebrations. Together they symbolized the presence of peace, order, and safety in the capital, and the firemen tended to march first in parades.[39] Spectators enjoyed seeing the ceremonial performances of marching troops carrying the tools of their trades, either guns or fire extinguishers, while wearing their shiny European-style helmets and pressed pants.[40] In his 1903 novel *Santa*, Federico Gamboa describes a lively Independence Day celebration. In the midst of the festivities, a company of firemen march four by four into the crowd carrying flags, banners, and torches to celebrate the event. Their presence awes the eponymous protagonist, the prostitute Santa, who describes the men as "mythological beings."[41] Parades such as the one described in *Santa* were common in various cities. In at least one instance, a volunteer brigade petitioned the municipal government requesting uniforms so that the volunteers could march in the Cinco de Mayo parade in Puebla. The Poblano volunteer firefighters wanted uniforms as a way to encourage the youth of the city to become active public servants.[42] At times, fire brigades rewarded civilians for their valiant efforts and to further encourage citizens to take on roles as public servants. The Veracruz fire department gave one citizen the title of "honorary firefighter" when he warned the fire brigade that the fire in a cigar workshop would inevitably cause nicotine poisoning and thus harm the men who tried to save the building.[43]

Popular opinion depicted firemen ambiguously, sometimes as heroes of the city, and other times as more destructive than helpful. After an 1895 electrical fire at a celebration for the Virgin of Guadalupe near the center of the city, a bystander explained how he had immediately informed the firefighting brigade, but, "like always," they had arrived late and offered no real assistance.[44] Tardiness was a common complaint about firemen. After a 1909 theater fire in Puebla, the front

Gran Cuerpo de Bomberos de Puebla.

FIGURE 3.3. "The Great Firefighting Brigade of Puebla" riding on the back of a giant tortoise to the scene of a raging fire. *El Dictamen*, vol. 7, no. 29, February 4, 1909, 1.

page of the newspaper *El Dictamen* had a cartoon of five firemen trotting to the scene of the fire on the back of a giant tortoise instead of a fire engine (figure 3.3).[45] The broadside and story about the Tepito fire describe firemen looting the goods inside the market rather than helping to extinguish the fire.[46] Another fireman, Herculano Benítet, was imprisoned for stealing a gold watch encrusted with diamonds while he should have been putting out a fire at a corner store.[47] Stories of dishonorable, violent, and disobedient firemen left some citizens worried, rather than relieved, when the fire brigade arrived to help.[48]

FIGURE 3.4. José Guadalupe Posada, "Corrido: La quemazón," n.d. Courtesy of Colección Andrés Blaisten.

A common dilemma found in the era of professionalization was that untrained bystanders felt ill equipped or even discouraged from assisting in matters better suited to the skills of experts. Delayed responses often had calamitous results. The time spent waiting for the fire brigade to arrive could mean the difference between a small, manageable fire and a large, unwieldy conflagration. A story from Charleston, Massachusetts, illustrates the unintended consequence of relinquishing all control and authority to experts:

> On a small house near the bridge, there was a little fire, perhaps three feet in diameter. An earnest man could have dashed it out in a minute or two, with a pail of water. No one, however, made an attempt to do what every one supposed would be done in a few minutes by the firemen. The firemen did not come as expected, and as there was a high wind, the fire quadrupled its proportions every minute, and soon the flames leaped upon two large buildings, which before the first engine got to work were all on fire. The people now saw their error, but it was too late.[49]

FIGURE 3.5. José Guadalupe Posada, "Fire at the Volador," n.d. In *Posada's Popular Mexican Prints*, 24.

Confusion and disorder at the site of fires persisted despite the presence of a professional brigade. Posada's cartoon "La Quemazón [The Burning]" vividly shows the chaos that could ensue during a fire. In the image, steady and confident firemen aim their hoses at burning buildings, but their calm demeanor does little to subdue the people in the smoldering marketplace. Vendors and residents, screaming and flailing their arms, have fallen into pandemonium. One woman's dress has caught fire, yet the firemen, still fixing their hoses at the buildings, fail to respond. Another man in the foreground looks badly burnt and the skin on his face appears to have melted and become disfigured. Unlike earlier images of fire-induced pandemonium, this particular woodcut depicts intense chaos alongside the orderly, professional, and uniformed firemen (figure 3.4).[50]

Another image of a fire at the Mercado del Volador, a market known for horrific fires during this period, shows firemen holding back crowds of people with bayonets in order to make room for the men to wheel in a large water pump (figure 3.5). Government officials intended the presence of firemen to eradicate such utter confusion in the face of fires, but Posada, the ultimate critic of social life, showed that a fire brigade did not mean a complete end to this public reaction to disasters. Its inability to control crowds led many foreign travelers to criticize the city's fire brigade.

CONTROLLING THE FLAMES—THE FIRE BRIGADE

FOREIGN CRITIQUES OF THE FIRE BRIGADE

On his visit to Mexico City in 1910, the editor of *Overland Monthly* observed the city's police and fire departments. He asserted that the police department was nearly perfect and one of the best in the world. Policemen underwent exhaustive training in gymnastics and sword fighting and were tested on the names and locations of city streets and alleys. *Overland Monthly* reported that so adept were the city's policemen that they could ensure that a woman could walk from one end of Mexico City to the other at one o'clock in the morning without being bothered by vagrants or robbers. Contrary to this praise for the police department, the editor explained that the fire department clung to "antiquated and foolish methods." While the editor never defined these methods, he did suggest that Mexico purchase US machinery to fight fires and hire US firemen to train the Mexican brigades.[51]

His assessment of the brigade mirrored the commentary found in a short fictional story printed in the *Mexican Herald*, a newspaper for US expatriates living in the capital. The short story is replete with tales of clumsy, dimwitted attempts by firemen to extinguish a mattress that had caught fire in a downtown apartment. After the firemen have taken out pieces of furniture and a parrot in his cage, they proceed to carry the owner of the apartment, Don Protasio, down three flights of stairs, despite his repeated protestations. When they reach the street the firemen accidentally throw him into an open sewage ditch where he drowns in a cesspool of fecal matter. When the firemen go back to retrieve Protasio's wife, it takes several firemen to hoist her over their shoulders and carry her down the stairs because of her portly figure. To fully extinguish the fire, the men spray an enormous stream of non-disinfected water into the previously hygienic apartment, giving the entire building a fetid odor.[52] This story mocked the capital's public services and made unflattering comparisons between Mexican and US firefighting techniques.

The foreigners' gaze was critical of Mexico City's fire department, and US observers thought that it needed the guidance and training of professional firemen from the United States. The editor of *Overland Monthly* argued that US firemen, recognized as the best in the world, could train and imbue excellence and efficiency in the lackluster Mexican brigade.[53] The description of US firemen as the best in the world was correct. At the 1900 Paris Olympics, volunteer and professional fire brigades from fifteen countries competed in an event known simply as "lifesaving" (*sauvetage*). Each brigade was timed on how quickly its members could run into a burning building, retrieve life-sized dummies, and put out the fire. The Kansas City fire company won the competition and journalists described the event as an exhilarating spectacle that elicited repeated and prolonged applause.[54]

Days before firefighters from around the world demonstrated their firefighting prowess in the lifesaving games, the Congrès International des Sapeurs-Pompiers (International Fire Brigade Congress) met in Paris to discuss best approaches to training and equipping fire brigades. Three delegates from Mexico—Colonel Rodrigo Valdes, Lieutenant Colonel Mauricio Beltran, and Lieutenant Colonel of Artillery Ygnacio Altamira—came to the meeting.[55] Other Latin American countries, including Ecuador, Nicaragua, and Peru, had interests in hearing about firefighting practices throughout the world and each country sent one delegate. The French, understandably, had four hundred officers present at the Congress. Representatives from Russia, France, and England, among others, detailed how their fire brigades received training and funding and offered suggestions for cities that were on the brink of creating brigades. The Russian delegate pointed out both the positive and negative aspects of having volunteer brigades, explaining that volunteer brigades "cost little or nothing," as opposed to the fifteen to thirty thousand francs it typically cost to establish a professional fire department. The delegate also noted that volunteer brigades "present serious flaws in terms of the speed of departure and maneuvering," and if possible, a professional brigade should be installed.[56] Mexican efforts to attend firefighting conferences and install a professional brigade did not mean that foreign observers let up on their criticisms.

Foreign observers in Mexico suspected that thorough training from US firefighters, alongside American-manufactured engines and equipment, offered a solution to the supposed sham that was the Mexico City fire brigade. Such descriptions of Mexican backwardness were common for the time. Foreigners criticized the capital officials' incompetence because they had not invested more time and money in the fire department. One observer compared municipal officials in the capital to authorities in San Francisco before the earthquake and fire that destroyed the city in 1906. He predicted that Mexico City would experience a massive fire, an eventuality for which officials were woefully underprepared. City council members knew that they did not have enough water, firefighting equipment, or firemen, which would cause severe devastation if a fire or earthquake comparable to San Francisco's struck the capital. Even though foreigners belittled public service initiatives, the city's officials had made conspicuous efforts to prevent and control fire hazards since the 1860s.

The story of firefighting in Mexico City offers just one example of the growing reliance on public servants to bring order to the capital. These fire experts emerged because hazards in the city had become more destructive, a trend found among all major urban centers at the turn of the century. Confronting urban hazards, many of which were brought about by modernization efforts, also required the modernization of safety practices. Mexico City officials kept abreast of the international debates and discussions about firefighting strategies and adopted the ones they thought fit the capital. The modern tools of firefighting—engines, hoses, helmets, and uniforms—equipped firemen with the aura of authority and the expertise to tame flames. A mild tension between the role of public servants and the place of technology in fire protection starts to brew in the next chapter about fire engineers. Later, these tools of technology will become available on the individual level to modern consumers who will be able to calm their own fears and ward off fire danger.

CHAPTER FOUR
ENGINEERING SAFETY

Doña Nicasia and her boarders had a superstitious respect for the engineer and his invention—And even some superstitious fear of the kind that illiterate people sometimes feel for books and words and numbers.

FEDERICO GAMBOA, *SANTA*

In 1870, engineers investigated one of the biggest fire mysteries in the capital's history. Residents awoke on the night of March 17 to the sound of ringing cathedral bells. The repeated clanging of the bells signaled that a fire had erupted and prompted people to run outside to see how they could help. Bright flames lit the night sky, illuminating the city and filling the air with smoke. One witness followed the light and smoke and found the reflections of tall and unruly flames dancing on the cathedral's walls, making the building look menacing, yet oddly picturesque.[1] Residents who had hurled themselves out of bed when the alarm sounded soon discovered that the Mercado del Volador, one of the city's principal markets, near the southeast corner of the Zócalo, had become engulfed in flames.

The flames entirely consumed the wooden stalls inside the market. The heat of the fire made glass windows and doors explode. After several hours, the violent blaze showed no signs of stopping, and onlookers feared the fire would spread to the nearby National Palace and the Metropolitan Cathedral. Volunteers removed all the wood and other flammable substances from shops and stalls in the surrounding streets to contain the fire within the walls of the market. The next day, when the fire had finally subsided, owners of fruit stands, hardware stores, and clothing boutiques returned to the market to find that their livelihoods had been reduced to ashes.[2] Market Inspector Juan García Brito lamented that the tragedy was an injustice imposed on the poor vendors of the city. The Volador fire displaced more than three thousand people, and according to García Brito, left them "without bread." He went on to explain that fire represented an urban misery that had come to infect the poorest classes like a cancer.[3]

In the aftermath of the fire, debris filled the city block that once held the bustling commercial venue. Within hours of putting out the flames, the Ayuntamiento appointed a commission of engineer-inspectors to begin an extensive investigation to uncover how the fire began, and more specifically, to determine culpability.[4] The commission settled on a two-pronged approach: first, an investigation of the scene of the fire to look for any clues of what could have caused the conflagration; second, a round of questioning vendors, night guardsmen, managers of the market, and other eyewitnesses. The process resembled a murder mystery in which everyone became a suspect and the inspectors publicly scrutinized the alibis and overall credibility of the witnesses.

In the immediate wake of disaster, people have a tenacious desire to pull together and help one another overcome the effects of devastation. Goodwill tends to fade as time passes and the disaster becomes less visible to those not directly affected. Following the Volador fire, neighborliness turned to ill will when the market fire morphed into a criminal investigation, and soon suspects started to point fingers and shift the blame onto other vendors or guards. Employees in the market recalled details of the day's events, noting any activity that seemed suspicious. One vendor vaguely remembered seeing a lit candle in the shop of Gregorio Chavarria.[5] Another merchant alleged that Gertrudis Sedeno kept a large supply of dry lumber inside her shop and that the day of the fire she may have lit an oil lamp.[6] One night guardsman accused his colleague, Plácido Sánchez, of accidentally breaking a pipe that could have supplied water to the plaza.[7] Another neighbor, after hearing what sounded like fireworks coming from the market, rushed to the scene, where he found a sleeping night guardsman at the gate.[8] The commission's investigation was filled with baseless speculation and anecdotes from witnesses, all of which helped divert attention and blame away from the witness being questioned.

In their testimonies, residents keenly dissected the actions of engineers and public works employees. One night guardsman was pleased to accept the help of engineer Carlos Moya and explained to the commission that considering Moya's esteemed profession and scientific knowledge, he expected the engineer would offer a great deal of help.[9] Other assessments of the impact of municipal engineers were less glowing. Training at the engineering school did not necessarily prepare men to take charge in times of chaos, and some witnesses suggested that the engineers' inadequate leadership abilities actually exacerbated the direness of the situation. Cayetano Gómez Pérez, who had rushed to the scene as soon as he heard the alarm from the church bells, noted that no one had spearheaded a fire extinction plan—not the director of public works, nor any among the number of engineers on site. He further explained that engineer Antonio Torres Torija had undertaken the task of preparing the portable fire pump held in front of the cathedral, but due to his feeble attempt to give orders and co-ordinate volunteers, the pump never arrived at the burning market.[10] While the commission of engineers tried to discover who started the fire and how it became unmanageable so quickly, they soon recognized that each engineer could have done much more to put out the flames at the Mercado del Volador.

Finding the culprit became a fruitless endeavor. The investigation revealed that the fire spread so fiercely due to a combination of forces, making it impossible to single out one person. Certainly, the negligence of Gertrudis Sedeno, who had stored a pile of lumber in her stall and possibly left an oil lamp lit, or that of Lázaro Gual, who had forgotten to blow out a candle, may have started the fire, but many other factors contributed to its relentless spread. The biggest problem, and the one that most of the eyewitnesses pointed out in their testimonies, was the insufficient supply of water at the market. Workers testified that the fountains in and around the market had dried up completely fifteen days prior to the fire.[11] The only source of water near the site was a neighbor's artesian well (a well with natural water pressure), but it lacked adequate pressure to shoot large amounts of water through hoses and onto the flames.[12] Despite consistent complaints about water, some witnesses, such as Ayuntamiento representative Manuel Patiño, stressed that the cause of the fire was a different matter entirely, and therefore the water issue should not be brought into the investigation.[13] Regardless, both residents and government officials refused to take all the blame.

The Mercado del Volador fire showed residents just how ill prepared the capital was for disasters of this scale. In its current state, no one could rely on Mexico City's water supply for emergencies. Reporters raised concerns about city markets succumbing to the same disaster as the Volador and suggested improving the existing city infrastruc-ture to prevent fire hazards.[14] Within two years the market suffered another fire, which more or less consumed all of the temporary stands that the Ayuntamiento had built after the 1870 conflagration, and yet again, it was a lack of water that exacerbated the situation.[15] A decade later, public health officials reported that the Volador's wells were still desperately in need of water.[16] Not until 1890 did the Ayuntamiento fully rebuild the market, a delay that indicated to residents that municipal authorities were not only unable to put out flames and control frightened crowds, they were also unable to help victims return to their businesses in a swift manner.[17]

Ensuring clear accountability in the future required some consensus about who in Mexican society possessed fire-related expertise. Within the context of late-nineteenth-century Mexico City, university-trained engineers were poised to become authorities in all things related to fire. The growing consensus that science and technology would improve daily life meant that, in theory, engineers could quantify risk by rooting fire prevention and control in scientific understanding. Mexico thus fell in line with a global trend that gave privilege to university-trained scientists and engineers, who increasingly were called upon to offer insight on how to confront problems that plagued cities. Yet much of the population ignored the experts' advice and instead trusted that the technology introduced by engineers could save the city from harm. In many ways, the presence and authority of engineers increased patterns of recklessness among residents who felt that the burden of fire safety had been lifted from the general public and now fell exclusively on trained engineers.

SCIENCE AND TECHNOLOGY IN MEXICAN SOCIETY

Science and technology, while seemingly objective, are rooted in historical context.[18] This is most visible in applied science and engineering, fields that directly respond to the needs of society. As in any other case, the particular milieu in which Porfirian engineers found themselves influenced the projects they chose and how they spent their time and resources.[19] This context was one in which government officials and urban citizens were turning to science as part of a modernizing moment. The rise of *científicos* (scientists) thus typifies this context. These unofficial advisers to President Porfirio Díaz had an abiding faith in science to solve national problems, even ones that had more of a political or social nature. Often trained abroad, they applied the principles of scientific rationality to planning and public policy. Water management, economic development, forest conservation, and fire prevention all came under the care of scientists and men with technical expertise. Politicians relied on the científicos to transform esoteric findings into applicable discoveries that could be used to solve the problems of every-day life.[20] It is important to make a distinction between científicos who advised on social issues and technical experts who planned and managed projects. While both were científicos, it was the technical experts, especially engineers, who were more involved in the everyday lives of capital residents.

Expanding the National Engineering Academy represented one of the ways in which government officials gave priority to scientific education. Their goal was to have citizens offer insights on how to improve economic development, while also addressing the pressing issues of city life.[21] During the Restored Republic period following the French occupation of Mexico, one of President Benito Juárez's first initiatives was to increase funding to the College of Mining (Colegio de Minería), in 1868. This transformed the institution from a colonial enterprise originally designed to ensure that precious metals would be exploited for shipment to Spain into the more comprehensive Expert School of Engineering (Escuela Especial de Ingenieros). This reflected the social and technological needs of a stabilizing period, when officials expanded their focus beyond rural, extractive, export-oriented economic activities and began to train engineers to diversify and improve domestic industries, often located in urban centers.

As a result of this initial period of expansion and professionalization, interest in engineering had grown to such a degree that by 1883 leaders had instituted a number of curricular reforms. They renamed the academy the National School of Engineering (Escuela Nacional de Ingeniería) and expanded its course offerings to give rise to cutting-edge occupations such as telephone engineers, railroad builders, surveyors, hydrologists, road and bridge designers, electricians, and sanitation experts.[22] In coursework and assignments, professors at the engineering academy taught students how to harness and control the energies and materials of the natural world for human benefit.[23] The trajectory of the curriculum directly reflected the developments that society had begun to experience: public access to electricity, city-wide sanitary restrictions, and indoor plumbing marked just some of those changes. Alongside their classroom training, students participated in a three-month practicum, authorized by the national government, in either railroad building, water methods, mineral expeditions, or mining.[24] The classroom and practicum settings not only trained students to be proficient engineers, they imbued them with a sense of civic duty. Purportedly, the National School of Engineering molded students into public servants who took on social and moral roles in society and applied their knowledge to improving city life.[25]

Unlike the privately funded universities in the United States, which could award engineering degrees without federal oversight, Mexican engineering academies participated in a nationally standardized process that resulted in a nationally accredited diploma.[26] Students accepted into the National Engineering School received free training made possible through public funds. A prominent engineer of the early twentieth century, José Ramón Ibarrola, in his assessment of engineering training, claimed that offering publicly funded educational programs compelled graduates to use their training in the public sector rather than for the purpose of accruing large salaries in private corporations, as students often did in the United States.[27] Francisco de Garay, an engineer of the time who was instrumental in designing Mexico City's drainage project, explained that engineering was not a lucrative profession, but a noble one that had intrinsic value in society.[28] Engineers thus carried an obligation to apply their understanding of science and technology to improve life, but also to put Mexico on par with other modern countries in terms of urban safety, infrastructure, and innovation.

Engineers crafted a professional identity associated with strenuous studying, thankless labor, and long hours, which helped them gain a reputation as a respectable, professional class of people who were willing to make sacrifices for the common good.[29] In the late nineteenth century, in Mexico as well as other Latin American nations, architects and engineers earned positions of power in cities and became recognized as experts in urban planning and safety.[30] Jorge Tamayo, a historian of the National Engineering School, has argued that the development of the National Preparatory School initiated an era of professional specialization and represented a major cultural turn, one that instantly made its graduates voices of authority in society.[31] On a daily basis, Mexico City residents encountered evidence of the engineers' presence and influence—signs explaining the maximum occupancy of buildings, scaffolding affixed to construction sites, and engineer-inspectors ambling the streets all served as reminders of the importance of this professional class. Centrally located near Alameda Park and the Teatro Nacional, the imposing National School of Engineering reminded residents that engineers were intrinsic to the material progress of the city. Furthermore, physical evidence of their technological prowess, including drainage canals and sewage grates, became markers of the engineers' craft.

The prestige associated with earning an engineering degree helped students to integrate into Mexico City's political system after they completed their studies. The engineering faculty held prominent positions in political life, giving them and their students high social esteem and reinforcing the idea that their work represented an advancement of the national interest. After his tenure at the engineering school, Blas Balcárcel became the minister of development (Ministro de Fomento) and actively took part in the expansive Porfirian drainage projects in the Valley of Mexico. Engineer Leandro Fernández, twice director of the school, trained hundreds of engineering students and also held the high-ranking position of secretary of communications and public works (Secretario de Comunicaciones y Obras Públicas). Antonio del Castillo, also a former director of the engineering school, rose to the position of head of the Geological Survey of Mexico (Comisión Geológica de México). The prestige of the engineering profession helped transform the government into one ruled by experts—or, simply put, a technocracy.[32]

Fire hazards, to one degree or another, became a concern in every branch of the engineering profession. Sparks from trains or improper coal storage practices on train cars often caused large conflagrations, forcing engineers who specialized in railroad construction to factor those risks into transportation designs. Electrical fires, a growing problem in workshops and theaters, required specialized knowledge from trained engineers to be put out safely and to be prevented in the future. When the existing hydraulic infrastructure could not provide sufficient water pressure to shoot water long distances out of fire hoses, hydraulic engineers built pumps and other apparatuses to overcome environmental shortcomings. Architectural engineers designed fire escapes, fire gaps, and firebricks to protect their buildings. Even sanitary engineers had to confront the fire problem by defining fires as public hygiene concerns and monitoring these potential hazards carefully. After the massive Volador fire, professional engineers trained in a number of areas took on influential roles as inspectors, using their knowledge to mitigate hazards in the city.

PREVENTING DESTRUCTION: FIRE INSPECTIONS

New occupations for professional engineers emerged as the need to confront the risk of fires increased. Engineers, valued for their education and ability to embrace the latest approaches in urban planning, were the ideal candidates for a number of new fire-related public positions. For example, the Ayuntamiento hired a city inspector to monitor potential fire hazards and to collect fines in each of the city's eight districts (*cuarteles*) and only considered applications from engineers.[33] After each fire, the Ayuntamiento called on the city engineer to evaluate the scene and to offer suggestions to the general inspector of the police department and the chief of the fire brigade about how to improve fire safety.[34] The official description of the position of fire chief of Mexico City called for a high educational level, noting that only a trained engineer or architect could fill the position.[35] Officials argued that engineers or architects would more fully comprehend the physics of building construction and the chemical reactions that influenced how fires spread.[36] Even in the legal sphere, engineers increasingly came to assist judges and attorneys in court cases, providing expert testimonies about the causes of fires and helping to determine whether or not foul play could be ruled out. Their efforts at investigating fire scenes helped form the basis of forensic science, a field that entered university curricula in the 1930s.[37] With a significant increase in the number and types of occupations that engineers fulfilled, these men became a far more visible presence in society at the same time that their social profile grew. Nevertheless, this visibility did not ensure universal admiration or respect. Rather, the engineers who came to occupy civic office received both praise and scorn from residents of Mexico City.

During particular years in which fires were noticeably more frequent, shopkeepers and businessmen alike complained to the Ayuntamiento that city engineers had not done enough to prevent fire hazards.[38] Neighbors on Tulipán Street explained to city engineers that the nearby La Fortuna match factory had ignited numerous times. In their letters, they asked that an inspector verify that the owner had followed safety protocols to prevent another fire from inflicting deadly consequences on the entire neighborhood. After uncovering a series of safety concerns, inspectors confirmed the neighbors' suspicions and forced the business to move to a less central location.[39] Residents continued to convey their worries through petitions to city engineers. In response to a devastating conflagration that left the famed luxury department store El Palacio de Hierro in ruin, downtown residents criticized city engineers, accusing them of being uninterested in public safety and hygiene. More than a month afterward, the streets remained blocked by scorched wood and crumbled asphalt, and numerous basements in the surrounding area remained flooded because firemen had doused the enflamed building with several thousand gallons of water.[40] The need for protection against fires required that engineers regulate safety. Despite their role as professionals who, by training, were meant to build things to serve human needs, they spent a lot of their time designing and enforcing regulations and policies. When it appeared that these trained experts had not enforced fire codes with enough rigor, residents quickly vented their frustrations. Conversely, when fire inspections uncovered negligence and disobedience of the fire code on the part of business owners or other citizens, engineers levied hefty fines and in certain cases closed businesses temporarily—measures that the business community disdained.

Especially strict safety regulations applied to theaters. These high-occupancy venues had gained a global reputation as tinderboxes and deathtraps. John Hexamer, a Philadelphia civil engineer and architect who specialized in insurance surveying, asserted, "Theatre fires can have but two eventualities: either the fire is extinguished in the first minute, or the entire theatre destroyed."[41] Gruesome stories of the 1873 Paris Opera fire, the 1903 Iroquois Theatre fire in Chicago, and the 1909 fire at the Teatro Guerrero in Puebla reminded engineers of the importance of regulating buildings with high densities of people, and none seemed to pack more people into tight spaces than theaters.[42] In 1878, German engineer August Foelsch published a statistical account of all of the theater fires (which were primarily in Western Europe and the United States) that had resulted in the complete destruction of the building. During the previous hundred years, 460 large theaters had burned to the ground. Foelsch noted that theaters in the United States had particularly short lives due to fires—on average, theaters there lasted only eleven to thirteen years.[43]

These venues ignited with such frequency due to unsupervised gas lighting, discarded embers from cigarettes, overheated stage bulbs, neglected electrical wiring, mismanaged film, and kindled stage props.[44] The vast majority of these fires occurred during performances, after hundreds of people had taken their seats and become so captivated by the show that they often failed to notice the signs of fire, which, according to Foelsch, tended to increase the death toll. Flames rarely acted as the principal cause of death; rather, smoke usually asphyxiated the victims before the flames touched their bodies. Several New York fire inspectors suggested creating flues in the roof or large skylights of glass over the stage that would allow the smoke to drift upward and thereby prevent suffocation.[45] In addition, during the chaos, mobs of panicked theatergoers would stampede their way through exits, trampling people to death in the process.[46] Felipe N. García, a witness at the fire that erupted at Mexico City's Teatro Alcazar, testified that the fire created pandemonium among the spectators. Frightened patrons scrambled around the aisles and jumped over seats to escape through the nearest available exit. García explained that it was nearly impossible to evacuate the building because hordes of people had completely blocked the stairwell and the owner had earlier designated the secondary exit as a storage area for extra chairs.[47]

Cinematic and stage theaters posed distinct fire problems. In playhouses, fires tended to start on the stages, which were usually constructed of wood and cluttered with props and backdrops. These items could easily fuel a fire started by an actor's cigarette butt or a spark from the stage lighting. Due to the location of the performance at the front of the theater, playhouses proved to be far less risky than cinemas: spectators could simply avoid the flames by exiting through the main entrance. Cinema fires, on the other hand, usually started in the projection booth, which sat directly above the audience and over the main entrance. Therefore, the site of ignition meant that debris often fell on top of the audience or flames made it difficult for spectators to escape through the main entrance.

The greatest factor that made movie theaters more prone to fire was the presence of nitrate film. The process of creating nitrate film began as early as the 1840s, when Christian Friedrich Schönbein, a German-Swiss chemist, created "guncotton," a combination of cotton cellulose and nitric acid, which he intended for use as gunpowder. Additional experiments that combined the nitrated cellulose with an alcohol revealed that when dried, the substance formed a transparent, flexible sheet, becoming not only the world's first plastic but also an accessible and affordable medium for motion picture film. The cinema world soon realized that nitrate film is highly flammable. Volatile in both warm and humid conditions, when it comes in contact with combustible organic materials such as wood or turpentine it can explode. Burning at a higher temperature than gasoline, its flames are more destructive than most fires. In addition to its high temperature, nitrate film emits lethal nitric acid fumes when burning that, if inhaled, can shut down the respiratory system. Perhaps the most troubling part about nitrate film is that once it ignites, the chemical reaction produces oxygen, meaning that ignited film can continue to burn while completely submerged under water.[48] Regardless of serious safety concerns, nitrate film remained the primary type of film used for motion pictures globally from the 1880s until the 1940s, when acetate film—at the time known as "safety film"—became the industry standard.[49]

Even the simple necessity of storing nitrate film presents challenges. The film gradually decomposes and turns into a flammable powder or slime, making its storage or transport risky. Archivists at the time found that storing the film at sustained high temperatures (over 100°F) and in airtight containers could create the conditions for nitrate film to combust spontaneously, whereas storing it at low temperatures in containers that allowed it to breathe and release gases could delay decomposition and prevent sudden fire outbreaks.[50] Despite advances in film preservation, for the most part the majority of these early films have been lost in studio storage fires. In 1982, Mexico's national repository of film caught fire due to nitrate film combustion, and more than six thousand films were lost.

In Mexico and throughout the world, a conundrum emerged: there was a growing population of moviegoers, but screening more movies brought more risks. Engineer-inspectors tried to mitigate these risks through frequent theater inspections, but not all business owners followed protocol, and the means for enforcing regulations proved inadequate. This was the case for the deadliest fire in the nation's history, which took place in a movie theater in the port city of Acapulco at the Teatro Flores. On Valentine's Day in 1909, the Teatro Flores celebrated its opening night, and thousands of spectators gathered to inaugurate the cinema, which was supposed to screen eight silent films. In this moment of excitement, theater employees overlooked the building's maximum occupancy guidelines and allowed many more spectators in than they should have. Hundreds more stood outside of the theater hoping to get a glimpse of the films. While audience members enjoyed the films, the heat of the projector bulbs caused the nitrate film to ignite. The projectionist attempted to smother the fire with his shirt, but the flames grew too rapidly and devoured everything in the booth in under a minute. The sound of screams from upstairs alerted the audience to the fire. Shortly thereafter, the projectionist, fully engulfed in flames, staggered down the stairs. With each step he took, the flames that blanketed the young man touched the wooden seats or the fabric that hung along the walls. Within minutes, the walls, screen, chairs, and ceiling caught fire. The crowd quickly rushed to the only exit in the theater. Its doors opened inward and with the frantic crowd pushing against them they proved impossible to open, leaving the audience trapped inside the burning theater. Before long the roof collapsed and fell on top of the crowd.

The Teatro Flores burned for six hours, spreading thick smoke throughout the city. Witnesses recalled that the smell of burning flesh remained in the air beyond the mountains for several days. In the days following the fire, prisoners dug a large pit, into which they shoveled the charred remains of bodies, and crowds assembled to identify loved ones. Still, most of the bodies were so badly scorched that most remains were unrecognizable. Approximately three hundred people burned to death that night. One particularly crass newspaper headline from Kentucky read, "Nearly Three Hundred People Caught Like Rats and Roasted."[51] The tragedy took a heavy toll on Don Matías Flores, the owner of the establishment, who felt personally responsible for the fire. In his attempt to open the theater on time, he had not ensured that even the minimum security measures had been met. For example, the city engineer had not inspected the theater, despite city regulations requiring such an action before theaters opened for business; there were no fire extinguishers, alarms, or sprinklers to protect the building; and none of the employees abided by the maximum occupancy regulations. The reality of the situation drove the owner to despair. Days after the fire, he committed suicide by shooting himself in the head.

Long before the Acapulco theater fire, public officials and engineers considered fires in theaters to be a serious threat that should be regulated accordingly. In 1870, on the recommendation of city engineers and the Commission for Public Recreation (Comisión de Diversiones Públicas), the police department issued a notice to theater owners mandating that they purchase and equip their establishments with water pumps to extinguish fires. The law gave them only fifteen days to purchase the requisite materials.[52] Lawmakers commended the regulation and expanded it to prisons, hospitals, schools, plants, and any public place where crowds gathered.[53] The new regulations had the potential to make public buildings much safer by offering protective equipment in the event of a fire. Regardless of the beneficial safety measures the regulation promised to bring, the law proved much harder to enforce than anticipated. One year after the pump requirement went into effect, engineer Antonio Torres Torija reported that only one theater, the Teatro Nacional, had a fire pump to protect its building.[54]

Discouraged by owners' unwillingness to cooperate with the new regulations, fire inspectors vigilantly enforced the rules and made examples out of delinquent theater owners. In 1878, in one of their routine inspections, engineers discovered a theater that had been erected with complete disregard for the fire regulations. After examining the theater, which had been built entirely of wood and lacked a fire pump and sufficient water, inspector Ángel González de la Torre predicted that a grave fire would inevitably erupt if the building remained in its flammable state.[55] After several failed attempts to contact the theater owner, Juan N. Cortina, demolition crews tore down the building.[56] One month later, after the spot on Santa Catarina Street had been left bare, Cortina finally responded to the inspector's letters. Interned as a military prisoner in Santiago Tlatelolco, Cortina had no way of leaving the prison to make the necessary repairs. Instead, he had contracted three men to tear down the wooden sides of the building and install water pumps for the facility. Knowing nothing about the recent demolition of his business, he apologized repeatedly for his delayed response and promised to resolve the situation within a matter of weeks.[57] This of course was an unusual case, but engineers often found that theater owners tried to skirt regulations and found it necessary to move forward with the demolition of the building.

Baffled by theater owners' negligence, fire inspectors questioned why anyone would refuse to implement mandatory fire laws that could potentially save a business in the event of a fire.[58] Engineer-inspectors, in an effort to stress the importance of fire codes, wrote a new building code in 1888 explicitly for the purpose of preventing theater fires. The new laws meant that city engineers and Ayuntamiento officials had to become intimately familiar with each theater in the city because all theater building plans first needed the approval of municipal officials before construction could begin. Inspectors acted as the liaisons between business owners and government officials by reviewing building plans to ensure that contractors used noncombustible materials, positioned all doors and windows to open to the outside, and constructed stairs from stone or brick rather than wood. Beyond these codes for the building itself, theater owners also had to equip their businesses with a number of fire control devices: various deposits of water (five cubic meters of water for small theaters that did not admit more than a thousand people, and ten cubic meters of water for bigger theaters), a fire pump inside the theater, and a telephone to call the fire department. Each infraction could result in a fine ranging from five to one hundred pesos.[59]

Engineers and businessmen had competing visions about safety regulations. In the view of the engineers, many of the easily preventable causes of fires persisted because businessmen refused to follow city-mandated codes. For businessmen, following all the regulations cost money and thus threatened the existence of their businesses. In 1905, the municipal government appointed a commission of engineers, architects, and firemen for the purpose of ensuring the public's safety in theaters and other entertainment centers. Owners of entertainment establishments continued to be the most egregious fire code offenders. At the turn of the century, city engineer Abraham Chávez made it his personal goal to prevent fires caused by negligent business owners.[60] After spending time assessing theater safety, he found that most theaters had an insufficient number of extinguishers, many of which were scattered in obscure locations, and that rarely did theater employees know where the extinguishers were or how to use them properly if an emergency should arise.[61] He established a team to conduct monthly inspections of all theaters to ensure that the owners complied with the fire regulations and the sanitary code. If owners did not meet the regulations, the inspectors withheld their business licenses and the theater could not show films or perform plays. By changing inspection protocol to include the examination of extinguishers, pipes, hoses, and water tanks, Chávez hoped that engineers would become well acquainted with how to use the newest machines and technologies of fire safety. Businessmen continued to lament that fire codes and regulations cut into their bottom line, and it seemed that the only way Chávez and his fellow inspectors could change business-owning practices was to inspect with more regularity and not let any venue fall through the bureaucratic cracks.

New regulations meant that engineers held an even more prominent role in determining how people conducted business in the city. The new laws required constant monitoring by engineers and the complete revamping of theaters.[62] The city engineer now determined how many spectators could attend a theater, and depending on that number, how many emergency exits must exist. Due to the presence of nitrate film, projection booths had strict guidelines: the room could only be made of cement, the reel containers had to be metal, and the door to the room needed to be made of iron or steel. Only two workers were allowed inside the room at one time and at no time were they permitted to smoke cigarettes. Additionally, the room needed to have several full buckets of water at all times.[63] Audience members had to adjust their movie- and play-watching habits by avoiding the temptation to smoke during performances, something that had been accepted as routine up to that point.[64] Theater owners and audience members alike scoffed at what they considered an unreasonable fire law that required windows in theaters to remain open during performances, complaining that the light from the window made the picture hard to see, and, during the winter months, created an uncomfortable draft.[65] The public had begun to discover that the mere presence of firemen and fire engineers did not mean immediate fire safety, and that in spite of their protestations, residents had to change their daily habits to improve the safety of their environment.

CONTROLLING NATURE: PROTECTION AND INFRASTRUCTURE

Engineers held strong convictions that through education and training they could control natural forces for human benefit.[66] Often citing British engineer Thomas Tredgold's 1828 definition of civil engineering, these engineers tried to combat the ills of the natural world through technology.[67] Tredgold, writing on behalf of the Institution of Civil Engineers in London, described engineering as the art of finding potential sources of wealth and energy in the physical world and adapting them for human use.[68] Harnessing water energy marked one of the most important engineering feats in Mexico City, a city beset by constant hydraulic challenges since the colonial era. During the late nineteenth century, simply supplying the growing urban population with water to drink, wash clothes, power industries, and put out fires proved to be a difficult task. Engineers found ways to divert streams, funnel liquid through underground pipes, and pump water out of previously untapped aquifers to provide residents with the water they needed in order to live in the city. Technological know-how helped engineers fix a major urban problem.

In a globalizing world, technology offered a comparative measure by which to assess national progress.[69] Through newspapers, advertisements, lithographs, and world's fairs reports, Mexicans encountered numerous opportunities to learn about technological innovations from around the world. This information gave them a way to juxtapose their nation's technologies with the machines and equipment found in other regions. In an 1897 report about new technologies, engineer Gilberto Crespo Martínez detailed the metric tons of coal, the length of railroad tracks, and the number of telephone poles in the United States, Great Britain, Germany, and France. He included this information as a way to compare Mexico with other countries, and he concluded by detailing the ways that Mexico might improve its technological shortcomings, citing formal education for all social classes as a pathway to national improvement.[70] Such comparisons often led to proposals for importing equipment or inventing similar devices. The belief that science was the key to advancement gave engineers a prominent social profile, since, in theory, they were the ones who could turn a society's technological and scientific aspirations into material progress.[71]

Natural water supplies had failed Mexico City's residents and could not keep up with the demands of the growing population, thus forcing engineers to design elaborate hydraulic networks.[72] Some of the water supply in Mexico City came from springwater from haciendas on the northern outskirts of the city, but most came from Desierto de Leones, Santa Fe, and Chapultepec. Although the springs offered accessible water to residents, hygienists believed this water was dirty, and none of these sources had much natural water pressure behind them, meaning the water could not be propelled long distances without technological assistance.[73] In order to use hydrants to extinguish fires, everyone living in surrounding streets had to shut off their water valves in order to increase water pressure, which meant that fire control required the public's cooperation and coordination. By 1872 the chief of police pleaded with the Ayuntamiento to have a map made of all the valves in homes and businesses so that policemen and guardsmen could more quickly fix the water pressure situation during fire emergencies.[74] Despite his request, the Ayuntamiento responded that it would be too difficult to commission an accurate map of all the valves, faucets, and wells located in homes and businesses throughout the city.

To begin the long process of improving water pressure to fire hydrants, engineers, beginning in 1870, spent seven years creating an extensive map of all the hydrants and corresponding pipes in the city. In addition to identifying the location of each hydrant, the engineers described the diameter of the pipe that supplied each hydrant with water. By 1877 more than fifty fire hydrants sat on street corners throughout the city (figure 4.1). Even the lower-class neighborhood of Tepito had hydrants, albeit only two to protect that entire area. For the most part, hydrants sat in front of government buildings, plazas, monuments, gardens, and some churches. In their reports on the locations and quality of fire hydrants, engineers noted that many of the hydrants near the periphery of the city did not work well. The hydrants outside of the city center functioned so poorly that the engineers thought it would be bet-

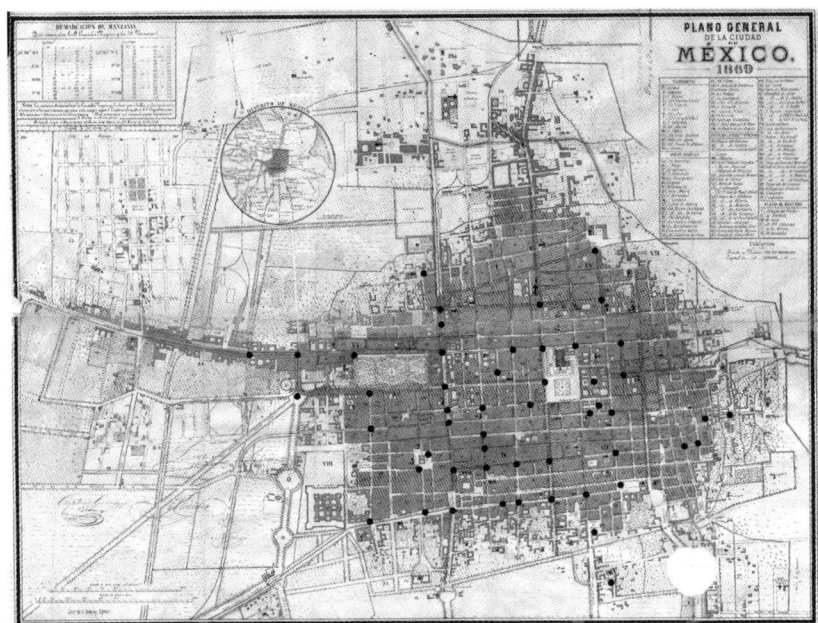

FIGURE 4.1. The black dots indicate the location of fire hydrants in 1877. Plano General de la Ciudad de México, Planero Horizontal 2, Gaveta 1, Plano 2, AHDF, 1869. For the engineering report that plotted the fire hydrants, see Se ordena al Ayuntamiento de la capital, remita a este Gobierno un plano de todas las cañerías de agua que cruzan las calles de la capital, para tomarla en caso de incendio, 1870–1877, AHDF: AGDF, Gobierno del Distrito, Aguas, vol. 1311, exp. 342.

ter to fill buckets of water at one of the nearby household wells than to attempt to use some of the hydrants for putting out fires.[75] Using this 1877 report and his own assessments of the capital's water infrastructure, engineer and businessman Carlos Medina discovered that of the 333 piped streets, only 185 had been equipped with the shutoff valves essential to fighting fires.[76] He estimated that the city lacked more than 1,400 valves needed to shut off water and put out fires quickly.

In their attempts to improve the city's water network, engineers encountered a series of issues that stemmed from previous infrastructural plans. Most importantly, they had to confront the earlier, uncoordinated and unregulated engineering efforts that continued to lie beneath the city. Ill-fitted lead pipes of various diameters meandered below the capital. Inspectors and engineers, appalled at the poor condition of the deteriorating infrastructure, were concerned about the

ENGINEERING SAFETY

leaky pipelines.[77] Leaks meant not only water loss but also that sewage and other materials could seep into the pipes and threaten public hygiene. In addition, the small trickles of water could easily break down the fragile lakebed soil that undergirded the city, which could be devastating to a city prone to earthquakes.[78] Each leak also decreased, if only slightly, the amount of water pressure in the pipes. Engineers tried to reconfigure and improve the water system but before doing so needed to repair or remove layers of old, intertwined pipes, establish distinct piping networks for potable water and nonpotable water (the latter being used for industry and fire extinction), and build a system that had sufficient water and pressure to put out fires.[79]

The cost of developing and improving hydraulic infrastructure in the capital was high, and the national and municipal governments ultimately spent roughly forty million pesos over a number of years to complete the projects.[80] During this expensive and time-consuming construction period, workers laid more than 2,500 tons of imported British pipes beneath the city. Storing water in elevated cisterns in Chapultepec Park allowed gravity to help to create the necessary water pressure to operate most faucets throughout the city. In Chapultepec Park, the highest point in the city, sat three British Worthington brand water pumps that sucked water up quickly (twelve cubic meters per minute) from large holding tanks and efficiently pushed it into the pipes that gridded the city.[81] While engineers had designed many of the pipes for drinking water or drainage, putting out fires remained at the forefront of hydraulic engineers' minds. In 1897 alone, nine high-pressure fire hydrants had been installed in the capital.

Foreign observers remained unimpressed by Mexico City's water infrastructure. In 1905 the Sanborn Fire Insurance Map Company sent mapmakers to survey the city and determine the potential for fires. Effectively, the entire city operated with two systems of water. The first came from Desierto de Leones, an open reservoir that sat in a forested area approximately twelve miles southwest of the city. Engineers built pumps to transfer the Desierto de Leones water to a concrete reservoir at Molino del Rey, only three miles from the city. Iron pipes were attached to this elevated reservoir and funneled water into the city, with nominal water pressure from the elevation. The second system was located at Chapultepec Castle and utilized springwater from the base of Chapultepec Hill. Again, pumps brought the water up to an elevated reservoir and funneled it into the city. Both systems could be connected by the seventy miles of pipes that sat below the surface of the city.

The Sanborn surveyors noted that neither system provided sufficient amounts of water or water pressure for their intended uses, and that residents often had to resort to using private wells for drinking and fire control purposes. The surveyors explained that many residents or businessmen owned small electric or hand-powered pumps that they could use to extract water quickly from wells.[82] The large, expensive, and visible infrastructural projects in the city still required that some residents fend for themselves and purchase personal technology for home use.

Even though engineers overhauled the capital's entire water network, fire protection technologies tended to benefit some sectors of society more than others. The fire chief expressed concerns that the majority of hydrants had been installed in strategic locations, often in the center of the city and in front of upper-class theaters or major business districts, thus assisting only the wealthiest sectors of society in emergency situations. Citing his experience with urban fires, the chief argued that the current fire hydrant infrastructure needed to be expanded drastically because fire hazards plagued all neighborhoods in the city, regardless of the socioeconomic status of their residents.[83] Engineers, in their attempts to make the city safer, knew that they could not provide fire safety selectively. They questioned the protection paradigm and used their expertise to argue for more equitable distribution of safety technologies.

With the new infrastructure came new regulations. City engineers assumed the task of inspecting hydrants and valves every month to ensure that they functioned properly.[84] Director of public works Luis G. de Ansorena claimed that a fire in a carpenter's workshop spread to three adjacent homes because the fire pumps had not been properly maintained and operated by men who knew how to work them.[85] In 1903 engineers recognized the importance of the monthly hydrant inspection after firemen tried to put out a fire in a corner store but could not use the hydrants because the pipes and valves had become clogged with dirt. Instead, firemen had to carry water, bucket by bucket, from a neighbor's artesian well, prolonging the disaster.[86] Soon thereafter, once engineers had conducted a series of repairs, the now revered hydrants became untouchable for anyone except trained professionals. Newly minted legislation that protected fire safety equipment stipulated that anyone caught touching or tampering with a hydrant would be fined twenty pesos.[87] Residents who had little or no access to water taps and fountains had grown to depend on the water in fire hydrants for household purposes, including bathing and washing clothes.[88] Every day for several months, a gardener who tended the plants in the Plazuela de Carlos IV used that plaza's fire hydrant to water the garden. When a fire erupted in the Plaza de Toros de Bucareli, firemen discovered that the gardener had somehow damaged it, leaving nothing more than a trickle of water to extinguish the fire.[89]

The need for water made engineers essential in the fight against fires in Mexico City. While drafting regulations and inspecting manufacturing workshops encompassed a major part of their jobs, it was through the technology and machines they created that they reminded people of the positive work they had done for the city. Hydrants, pipes, faucets, and pumps visibly marked the city with the influence of the engineers' craft. The actions of city fire engineers earned recognition not only in Mexico City but also on a national level.

THE DOS BOCAS PETROLEUM FIRE

Stories of engineers' fire expertise went beyond the confines of the capital. In 1908, engineers who had been trained in Mexico City were hired to put out a months-long petroleum fire in Dos Bocas, Veracruz. Extinguishing a rural petroleum fire would be a new challenge for engineers who had been accustomed to handling smaller urban fires. Representatives from the British-based oil company Pearson and Sons Ltd. had discovered a massive reserve of petroleum in the Dos Bocas region near the Gulf of Mexico. In the process of drilling a seven-hundred-foot well, workers unlocked an extraordinary amount of pressure that released more than thirty tons of petroleum. The burst of oil immediately came into contact with the flame from the boiler that powered the derrick, and the newly excavated well exploded, sending a geyser of fire more than a quarter mile into the air.[90] Nearly two months of trial-and-error efforts to extinguish the fire, compounded with extensive destruction of tropical forests and animal habitats, led people at the local and national level to develop a growing suspicion about the petroleum industry and its attitude toward safety regulations. Once again a fear of technology—in this case of petroleum-excavating technology—drove concerned citizens to question whether new machinery had pushed beyond the limits of safety.

The fire was a cataclysmic event for everyone involved—Huasteca residents, firemen, trench diggers, doctors, Pearson employees, nearby ranchers, and especially the flora and fauna of the area. Reports that documented the tragedy noted that shortly after the fire began workers felt the effects of the smoke, particles of petroleum, and other noxious gases in the air. Workers, firemen, and residents of nearby villages experienced stomachaches, headaches, swollen gums, and difficulty breathing. Some died instantly of asphyxia or burns.[91] More than a dozen firemen became so ill that they could no longer stand.[92] For the most part, the area was relatively uninhabited by humans and consequently only a few people died as a result of the fire.

Despite the relatively low death count, the effects of the conflagration extended beyond humans, annihilating the environment of the region. The oil flowed with such intensity that it reached nearby Lake Tamiahua and covered the water with a film of crude oil. As the toxic air poisoned the forest's birds and covered its foliage with ash, the marshes and lakes became overcrowded with the decomposing corpses of fish, shrimp, lizards, and turtles. The fire engulfed forty acres of tropical forest and the smoke and ash forced birds to migrate to safer environments. Even further inland, hundreds of head of cattle died from the fatal levels of sulfur, oil, and other lethal materials that infiltrated their water supply.

While some people lamented the destruction of this seemingly Edenic land, company officials expressed concerns about what the fire had done to their bottom line. After a month of constant fire, the well, which had begun as a small, fifteen-centimeter-diameter hole, had become a four-hundred-foot crater that spewed more than one hundred thousand barrels of valuable black gold every twenty-four hours, with an estimated loss of fifty thousand pesos each day.[93] Pearson representatives bemoaned this loss and speculated that Dos Bocas could have been the most profitable well in the world, outpacing the current top-producing single wells (Russia's most profitable well produced fifty thousand barrels each day, while only nineteen thousand barrels flowed daily from its largest counterpart in the United States).[94] Engineer Carlos F. Ganahl agreed that this was the richest well he had ever seen but countered that statement by saying that the fire was the most terrible of its kind and hoped that no man would ever have to combat such a conflagration again. Firemen and workers compared their experiences to the devastation that had occurred at the San Francisco earthquake and fire of 1906 or the Pennsylvania Oil Company fire of 1880 in Titusville. Engineers who had firsthand experience with fire hazards wondered if the risk could be decreased in the future with further regulation and on-site fire safety technologies, advocating the same type of preventative solutions that they and other public officials had prescribed to Mexico City.

This monumental disaster required immediate attention from skilled engineers trained in fire prevention. Mexico City engineers had gained a reputation for being adept at solving problems associated with fire hazards, which led the Mexican Executive Cabinet, in cooperation with the Office of Public Instruction (Despacho de Instrucción Pública), to send several professional engineers and students from the Mexico City engineering school to assess the situation in Veracruz. For this scientific excursion, the treasury department gave the team of engineers 550 pesos to put out the fire and take photographs and film of their attempts.[95] Manuel Villaseñor, Dehesa Benítez, Eleazar Núñez, Alberto E. Castillegjos, José Treviño Garcia, Ignacio Medina, Carlos F. Ganahl, and Emiliano Martínez used the expertise they acquired through their training at the National Engineering School to attempt to extinguish the column of fire spewing from the earth.[96] Regardless of the fact that their experience had been limited to handling urban fires that usually had started in factories and homes, the presence of the Mexico City engineers gave the British company and Huastecas hope that the fire could be put out.

Tackling the fire took many attempts. Some engineers from Mexico City, who were not part of the official team sent to Veracruz, sent in their theories and proposals for attacking the flames without ever seeing its extent with their own eyes. Initially, men on the scene used fire equipment that had been intended for city fires. They emptied several thousands of gallons of water from six British Worthington fire pumps directly on top of the flames. When this proved completely unsuccessful, more elaborate design plans emerged. One engineer designed a trench that surrounded the crater, in an attempt to collect excess petroleum and route the oil away from the swamp. Four hundred and ninety-five soldiers of the Sapper Battalion (Batallón de Zapadores) worked day and night digging the trench, which initially succeeded in diverting petroleum, but the company did not have enough holding tanks and therefore diverted only a fraction of the petroleum that they had anticipated.[97] Another plan sought to cover the wellhead with a large, heavy metal cap in an effort to deoxygenate the fire and thus extinguish the flames.[98] Engineer Manuel Villaseñor wanted to spray the fire with carbonic acid and water, a mixture typically found in handheld fire extinguishers.[99] The attempt that gained the most attention in newspapers involved sending two thousand tons of gravel and sand into centrifugal pumps that shot the rocks and debris from a substantial distance into the blazing well.[100] It took days for workers to find enough gravel and debris to shoot into the well, and hour by hour more and more people became ill from the smoke.

When the Mexican government announced that it would give two hundred thousand pesos to anyone capable of extinguishing the fire, Dos Bocas gained substantial international attention, and suggestions as to how to confront the fire poured in from around the world.[101] In an interview with *El Imparcial*, an engineer from the United States who had helped put out a similar fire for Standard Oil suggested drilling a second, parallel well several hundred meters away, placing a torpedo at the bottom of the well, and then shooting it horizontally in the direction of Dos Bocas. The torpedo method would create a deep passageway from the new, empty well to the one on fire, allowing the petroleum from Dos Bocas to funnel into the empty space.[102] Despite this sounding reasonable at the time—and it having worked in a similar fire in the United States—workers and engineers who were aware of Veracruz geography soon realized that the new well was at sea level, and thus the torpedo would cause oil to seep into the ocean. Engineers had designed all of these proposals to save at least some oil from the rich well.

Neither science nor machinery nor human inventiveness could put the Dos Bocas fire out. More than fifty days after the well caught fire, its fuel source exhausted itself and the fire ended along with it. With the horrific fire behind everyone, Juan Palacios, the head of the engineering expedition, openly pondered whether petroleum was worth the risk. He commented that while accidents occasionally happen, it would always be unclear if Pearson and Sons had done everything in its power to prevent such a tragedy. In 1908 the growing petroleum industry seemed as if it had no limits. Foreign and Mexican businessmen alike had profited immensely from exploiting the country's rich subsoil wealth. But after the two-month fire, Palacios and other onlookers expressed concerns about how petroleum excavations would affect the health and safety of all citizens involved in extracting and transporting Mexican petroleum. While this one had erupted in a place with a small population, engineers wondered if something similar could happen in cities and increase the human death toll. Palacios also wondered why Mexicans were willing to incur such high risks when other, less dangerous fuels, such as coal and charcoal, were readily available.[103]

Different individuals and groups reached different conclusions about the event. For the Huastecos who lived the nightmare, it was a tragic and emotional experience that reminded them of how dangerous and brutal nature could be.[104] For those critical of foreigners extracting Mexican resources, it was yet another symbol of Mexico's longstanding domination by imperialist powers, and a harbinger of bad things to come. Yet for businessmen and foreign investors, it was an anomalous and relatively meaningless isolated occurrence. Despite differing opinions, the fire showed the rest of the world that the Huasteca held rich oil fields ripe for the taking.[105] In disregarding the hardships that the communities experienced, these interpretations symbolized the stubborn and arrogant drive for modernization at any cost, an agenda that paid lip service to safety when profits were on the line.

For Huastecos, the Dos Bocas saga ended in tragedy and represents another relationship between class and disasters. But for Mexico City engineers, the event showed that they had earned national appeal. The national government depended on their expertise, albeit in a completely different spatial setting with vastly different fuel sources, to assist with a major natural disaster that threatened the lives, property, and health of thousands of people. Their status as professionals made them instrumental in putting out a seemingly uncontrollable fire, or at least trying every conceivable method to do so.

In the face of a new and dangerous fire regime, disaster expertise became essential to the development of Mexico City. Residents demanded explanations for the changing character of fire to help them understand why their city had become more hazardous. In a period when residents and officials privileged scientific and technological reasoning, engineers were the ideal candidates to explain the changing nature of fire. Federico Gamboa's fictional account of Mexico City at the turn of the century detailed this fascination with science and technology in his discussion of the engineer-inventor Bruno Ripoll. The other characters in the novel are in awe of Ripoll and his technological and scientific knowledge, so much so that they seek out his advice about daily matters, whereas previously they had asked a Catholic priest.[106] In the boardinghouse where Ripoll and the other characters live, his claim to fame is that he is inventing a submarine. For years he draws sketches and tinkers with model submarines, but his idea and his invention never materialize. It is both the promise of technology and science and the incomprehensibility of how the submarine works that make the other characters admire Ripoll. Similarly, the promises of technology and science for daily life made Mexican officials, businessmen, and residents look to engineers for answers.

Under the authoritarian, modernizing regime of Porfirio Díaz, Mexico in the late nineteenth century entered a moment of political and economic stability. For the first time since its independence a half century earlier, the government did not have to confront some combination of foreign invasions and civil wars. Thus, it became more capable and responsive to the needs of the society and its citizens. High regard for science and technology informed the way that officials made decisions and exercised their authority. Ultimately, this moved the government to embrace a technocratic form of political administration led by experts. This was visible in the rise of the progress-minded científicos, which included the engineers who regulated and modified the built environment. Yet engineers' roles in civic life were not limited to the highest levels of government. Their education and social prominence transformed them into public servants who regulated and inspected city life to ensure safety.

Restricting the examination of science and technology to the professional engineers and public servants bypasses an opportunity to see the fuller picture of fire safety in Mexico City. The prominent social roles of engineers and científicos, combined with the buoyant economic atmosphere of the late nineteenth century, inspired lay inventors and small-scale entrepreneurs to build machines for human benefit. While up to this point there has been an emphasis on the municipal government's responsibility for fire safety, there was also a competing trend that aligned with liberal notions of individual responsibility and self-reliance to prevent or control fire. In the parallel processes of public and private responsibility, technology was the thread that bound them together. The next chapter moves the analytical lens from citywide ordinances and infrastructure to household technologies by examining the initiatives of untrained, entrepreneurial inventors, who provided the capital with innovations in safety against fire hazards.

CHAPTER FIVE
INVENTING PROTECTION

The other boarders began to regard both inventor and invention with reverence, always speaking softly when at home and taking his every pronouncement as scripture . . . proud to live so close to a true genius who was bringing life to a fabulous creation, a submarine, that belonged somehow to the entire boarding house and that they therefore embraced with fierce adoration, even in its current state—tiny, imperfect, unfinished.

FEDERICO GAMBOA, *SANTA*

In the last decade of the nineteenth century, the capital required more protection against fires than ever before, and public servants helped provide some of that protection. By 1900 Mexico City had experienced prosperous commercial development. It housed nearly one hundred clothing and drug stores, more than two hundred lawyers' offices, four major banks, and a host of warehouses, real estate offices, telephone companies, markets for foodstuffs, and chemical suppliers.[1] With greater output and investments came greater risk of loss. The dangerous urban environment forced business and homeowners to assess and manage financial risks in a way they had never needed to do before, leading them to develop safety measures on their own rather than relying on public officials to do it for them.

Assessing only public sector occupations such as firefighting, fire inspection, and fire-related engineering ignores the important contributions that other citizens made to the effort to combat fires. Some capital residents saw business opportunities in fire hazards. A generation of inventor-entrepreneurs capitalized on growing fears of fire and sold products that protected people and investments from fire. Yet the process of exploiting disastrous situations was anything but a new phenomenon. In their analysis of the aftermath of disasters, several scholars have exposed the tendency of certain groups to exploit catastrophe and promote capitalist interests by advocating that private capital, rather than public funds, should be responsible for responding to disaster.[2] In nineteenth-century Mexico, businessmen who sold fire safety technologies advertised their products to people who had fresh memories of hardship and loss, who were therefore eager to prevent future catastrophe by paying for individualized protection. Their products offered customers a future commitment to protection and a way to prepare for uncertainty, whether that protection came in the form of an automatic smoke alarm to alert people to danger or a handheld extinguisher to smother flames.

While historiographical interpretations of this time period are full of examples of technology transfer and Mexicans importing and adopting technology, medicine, and equipment from the United States and Europe,[3] the important homegrown attempts at technological improvements and business ventures remain understudied. Environmental and business historians of Latin America have traditionally emphasized the region's export-based economies to understand underdevelopment as byproducts of colonialism and dependency. Yet this approach discounts the importance of local, non-export-oriented enterprises. Currently, most research on local economies in Latin America at the turn of the twentieth century has shown that cultures of entrepreneurialism tended to come from European and American immigrants who settled there.[4] Inspired in part by foreign fire safety technologies, Mexican businesses that sold protection against fire emerged.

EXPOSITIONS, DEMONSTRATIONS, AND THE DISPLAY OF TECHNOLOGY

The exchange of innovations from one society to another has been the cornerstone of technological development for centuries.[5] While many studies of technology and inventions highlight the role of certain nations in driving innovation,[6] in the case of late-nineteenth-century Mexico City, the influx of technology from around the world inspired people to invent locally. Machines on display at international expositions, discussions of technology in newspapers, and images of gadgets and devices in advertisements all allowed individuals to compare their technological progress with the advances being made in other countries. Local inventors challenged the idea that technology had to be imported from the United States or Europe. Instead, they used knowledge of local conditions to build machines and gadgets that benefited people living in the capital. The national and local governments also had a vested interest in promoting local innovation. In 1874, the Ayuntamiento invited citizens and resident foreigners to participate in the municipal exhibition, which celebrated the technological, gastronomic, intellectual, and scientific achievements of the country. Modeled on the Great Exhibition of the Works of Industry of All Nations of 1851 (called the Crystal Palace Exhibition) held in London and subsequent exhibitions in France (1855 and 1867) and Austria (1873), officials promoted the municipal exposition as a way to celebrate the nation's progress while serving as a practice run for the upcoming Centennial Exposition in Philadelphia to be held in 1876.[7]

The municipal exhibition's promotional poster epitomizes how officials wanted the country to appear to outsiders (figure 5.1). By collapsing time and space, symbols of past, present, and future wonders of the nation could exist within the confines of a single image. An eagle, a national symbol found on the country's flag, hovers over objects of material wealth while holding between his claws a sign announcing the exhibition. From left to right the print reads as a pictorial epic of the country's history. Pre-Hispanic artifacts propped up by flora and fauna describe the fecundity and wealth to be found on top of and beneath the soil. These symbols of the past blend into scenes of musical instruments, painters' palettes, and scrolls or bound books, which represent the nation's legacy of artistic achievements. The right half of the drawing portrays the present and future, represented by the tools of technical experts—large industrial cisterns connected to billowing smoke stacks, an assortment of cogs, wheels, anvils, levers, and scattered measuring instruments. A woman—either la Patria or Minerva—in a flowing white gown, her head adorned with a laurel wreath, sits among large industrial equipment, admiring a machinist's hammer while bracing herself on a large metal cog. This poster conflated the progress associated with machines, tools, and art with the iconography of the nation, evoking sentiments of patriotism and calling people to rise up and achieve their artistic and scientific aspirations at the citywide exhibition.

The potential for wide public recognition provided a powerful incentive for inventors to submit their creations or display their designs at the municipal exhibition. In each of the categories—agriculture, mining, manufacturing, industrial machinery, and art—the committee awarded gold, silver, and bronze medals.[8] Genaro Vergara won first place in the mechanics category for his machine to harvest tobacco; José Refugio Terraza won first prize in the clothing category for his method to mass-produce fine cloth of different colors; and Miguel González,

FIGURE 5.1. The poster from the Municipal Exhibition of 1874. Reglamento Municipal Exposición, November 1874, AHDF, Exposiciones, vol. 1037, exp. 40, caja 1, carpeta 28.

Alfredo Garay, and Luis González won first place in the naval category for their design of an indestructible safe to transport money on seafaring vessels.[9] Although these awards did not include monetary prizes, inventors hoped to earn public recognition by having their names published in the exposition's pamphlet and receiving their awards in a public ceremony. To further entice inventors to submit their work, the commission convinced the Ayuntamiento to allow a three-month-long patent-free period for anyone who participated in the event. This meant that anyone who displayed agricultural equipment, fire extinguishers, or industrial machinery could freely sell the products in the capital for three months following the exposition without paying burdensome licensing fees to receive a patent. This decision encouraged anyone with an invention, whether it was a machine, idea, or substance, to participate in the exhibition.

In many ways, these smaller municipal and national exhibitions functioned as dress rehearsals for larger international fairs. The opportunity to exchange ideas, see new technologies, and understand other countries more fully helped to increase interest in international exhibitions and fueled inventors' creative and competitive spirits. In a letter to the Mexico City Ayuntamiento inviting residents of the capital to participate in the 1911 International Municipal Congress and Exposition held in Chicago, the organizing committee explained, "It is believed that through this municipal exposition, making possible comparisons, one community with another, of their systems of government, their notable accomplishments of the past and their plans for the future, a spirit of rivalry and civic pride will be developed that will do much for the advancement of municipal betterment."[10] One of the avenues to achieve "municipal betterment" was to visit the large number of exhibits about urban police and firemen, which highlighted new designs in fire engines, fire alarms, fireproof vaults, fire escapes, fire extinguishers, and firemen's uniforms. Ayuntamiento officials certainly agreed that the Chicago exhibition was a worthwhile opportunity to learn how other municipalities approached similar problems that Mexico faced, and they agreed to send a representative from Mexico City.[11]

Most residents of the capital could not attend international expositions, but they could read about the new inventions displayed at these fairs. Advertisements, articles, and pictures of inventions led residents to contemplate the endless possibilities of new technologies for human benefit. News of technological advancements, especially those that could improve fire safety, bombarded the city. Newspapers carried articles about French chemists who developed substances to put out chimney fires, British firefighters who extinguished fires with high-powered mechanical pumps, and Parisian inventors who developed steam-powered fire engines.[12] Testimonies from fire departments in Cleveland, Cincinnati, and Washington, DC, translated into Spanish and printed in Mexico, told stories of the usefulness of fire devices like fire alarms and extinguishers.[13] All of these inventions promised amazing results to create safer cities.

Some of the public interest stories included animated eyewitness accounts of machinery or chemicals used to put out fires, which prompted government officials to import these technologies for municipal use. José Fernández, Mexico's minister to the United States, impressed by a demonstration of a chemical used to put out fires, wrote a lengthy account of the new product and sent it to the Mexico City Ayuntamiento. While living as a diplomat in Washington, DC, Fernández witnessed agents from the Harden Fire Grenade Company conduct an exhilarating demonstration of their product's effectiveness in putting out fires. The agents constructed a wooden wall in the middle of an open plaza and doused it with tar, petroleum, and turpentine, scattered twigs and dried leaves along the base, and lit the wall on fire with a match. The wall ignited immediately. As the flames grew taller, nearby shop owners and residents hurried to the scene to watch the spectacle unfold. The growing crowd of people felt the heat of the flames on their skin and had difficulty breathing in the heavy smoke it produced. Once the fire had sufficiently spread, a Harden agent emptied four handheld bottles of his company's chemical substance onto the flames, extinguishing them completely. Fernández included testimonies of foreign experts ranging from a US fire chief to British fire insurance agents to show how other major metropolises had already adopted these new safety technologies, warning that Mexico was woefully underprepared to protect its citizens from fire.[14] Only a few weeks later, Charles M. Martin, another Harden agent, petitioned the Ayuntamiento, asking for permission to conduct the same public demonstration he had done in Washington, a request that the Ayuntamiento enthusiastically granted (although no such demonstration occurred for two decades).[15] The Ayuntamiento eventually purchased sixteen boxes of Harden extinguishers intended for public buildings throughout the city.[16]

While public fire grenade demonstrations had become commonplace in the United States, England, and France during the last half of the nineteenth century, until 1902 Mexicans had only read about them in newspapers.[17] On the afternoon of July 24, 1902, the Harden Fire Grenade Company treated Mexico City residents to a thrilling firefighting exhibition. Just as representatives had done in Washington, in Mexico City they coordinated a public demonstration in one of the city's main plazas, built a large wooden wall, covered it in flammable substances, and lit it on fire. Ten minutes after igniting the wall, the Harden agents hurled three thin-walled, blue glass balls filled with a salt water and sodium bicarbonate solution onto the blaze. The glass balls shattered against the wall, allowing the liquid to burst out. Sodium bicarbonate's deoxygenating effect deprived the fire of one of the elements it needed to survive, while salt water dispersed the solution effectively throughout the area. The audience, accustomed to fires being put out slowly by volunteers carrying buckets of water, roared with excitement at the sight of three small handheld fire grenades destroying every last glint of flame.[18]

Several days later, the Harden representatives repeated the performance for a high-profile audience that included the secretary of public works, the chief of the fire brigade, the president of the municipal government, and the inspector general of the police department.[19] City officials all agreed that the grenades were ideal for fire-prone Mexico City. Because of the grenades' simple design, they boasted, even children and maids could use the glass ball to put out fires with ease. One bystander suggested that residents would no longer have to purchase costly fire insurance. Instead, the city could be equipped with grenades to ward off destructive fires.[20] Days after the 1902 Harden demonstration, journalists reported that in France and most European nations nearly every public building, hospital, school, estate, and private home housed at least one fire grenade. Such comparisons often led to proposals for importing equipment or designing similar safety devices.

A common thread throughout the stories of new inventions was the affirmation that Mexicans could easily invent better technologies than those found elsewhere. In 1901 the *Boletín Municipal* published an article about electric machines for firemen. Among other things, the author surveyed the country's interest in fire safety, and he mentioned some of the developments that surfaced in the wake of major fires like the one at the Mercado del Volador. He also summarized an article that he had read from *Revue Universelle*, a French magazine, that described an automobile with an electric pump that was praised as far faster and more reliable than the horse- or mule-drawn cars employed by most firefighting brigades. This fire pump, first admired at

FIGURE 5.2. Harden agents demonstrating how to use the fire grenades in front of a crowd in Mexico City. "Las granadas 'Harden' contra incendio: un invento maravilloso," *El Mundo Ilustrado*, August 3, 1902.

the 1900 Paris world's fair, generated international attention from city engineers, urban planners, and firemen. Rather than suggesting that the Ayuntamiento order this particular piece of equipment that had awed audiences at the world's fair, he unequivocally stated that Mexico's intelligent and creative engineers and inventors could develop a better product than anything that could be found in the United States or Europe.[21] One way to prove this was to patent Mexican inventions for recognition in Mexico and abroad.

PATENTING SAFETY

Patents, a seventeenth-century development, provided the legal procedures to protect intellectual property. Without them, many inventors feared that their ideas would be copied before they had an opportunity to profit from their designs.[22] In applying for a patent, inventors received the security of temporary monopolies (ranging from five to fifteen years) on their inventions. This process of protecting ownership of one's ideas facilitated capitalism, promoted a capitalist ethic in science, and offered profits as an incentive to conceive of new ideas and designs. In 1832 government officials modified an earlier Spanish policy to create the first national patent law. Despite slight modifications demanding that inventors supply exact descriptions, drawings, and models to the patenting office before approval, inventors complained that the laws appeared convoluted, contained vague language, required high fees, and had unclear adjudication procedures. Legislators rewrote the law in 1890, extending the patent term from ten to twenty years, decreasing the cost from upwards of three hundred pesos to 150 pesos, and changing the types of patents from only registering new inventions to accepting any discovery, improvement, method, or application intended to improve industrial development.[23] Soon applications for everything from farming equipment to bicycles to fire extinguishers stacked up in the office of the Ministry of Development, and each authorized patent received public notice in the *Diario Oficial.*[24]

The formal processes to register inventions required that individuals complete a series of official steps to profit from their ideas. Submitting a patent request could lead to months of correspondence with government bureaucrats who wanted clarifications, additional designs and drawings, or more elaborate details about its function in society. Inventors had to fill out paperwork and submit formal requests that justified their work and demonstrated its innovative qualities. Once the Ministry of Development deemed an invention, chemical, process, or idea a new innovation, the request had to await presidential approval. The last step in the process required payment of fifty to one hundred pesos, a cost that deterred some inventors from applying.[25] Some citizens felt that the formal patent process stifled the creative process and discouraged people from pursuing these enterprises, an outcome that ran contrary to its intended purpose of motivating inventiveness through protective incentives.

Copyrights and patents on intellectual property conferred high esteem on inventors and rewarded them for their inventions. By demonstrating their intellectual merit through technological innovation, inventors had the right to receive all the benefits that came with that invention, whether the prize was monetary compensation or social praise or both.[26] Patents armed inventors with legal recourse against anyone who borrowed the design of a particular invention, and such conflicts appeared with increased frequency in the judicial system.[27] In 1885, for instance, a heated legal dispute over Enrique Hernández Aranda's patent application for safety matches arose when E. M. Arzac and Company claimed that it owned the rights to the particular combination of chemicals used by Hernández Aranda.[28]

Patent law stated that discoveries, inventions, and new processes could be protected, but it also encouraged others to improve upon existing innovations. In theory, patenting ensured that technological innovation did not stagnate but rather that people constantly tried to better the invention for social good. Individual initiatives to improve life in the city presented itself as a profitable endeavor. Yet the inability to obtain credit discouraged inventors from going through the lengthy patent process. Lack of funds debilitated potential entrepreneurs and inventors, especially those who had grandiose ideas that would require significant capital. For some, the risk of venturing out alone without sufficient credit opportunities proved too great.[29] Some inventors even sought out investors in classified ads. One example of such an ad is from an inventor who had created a machine to quickly make masa dough for tortillas but did not have the capital to start a business and profit from the invention and thus had to solicit funds.[30] A fictional tale published in the *Mexican Herald* recounted the story of a downtrodden inventor who had complete confidence that his invention would bring him immediate success and unending financial freedom, but without investor interest or access to a bank loan he resorted to stealing a thousand pesos from a wealthy widow.[31] Without willing investors or the ability to obtain credit, it was difficult for Mexican inventors to establish themselves as technical experts. Instead, foreign inventors tended to profit the most in Mexico.

Rather than investing in and cultivating local expertise, the Díaz administration frequently looked abroad for technological solutions. By reforming the constitution to encourage foreign investments, Díaz and his advisors lured foreigners to Mexico. Free land, cash incentives, and reduced taxes motivated foreign investment. The presence of successful foreign businessmen, Díaz and his administration thought, would teach Mexicans to manage their businesses and affairs properly to generate greater profits. Additionally, the Díaz administration loosened its patent restrictions for foreigners as a way to increase their investment prospects and to create a hospitable business environment.[32] In 1890 legislators revised the patent law to make it easier for foreigners to bring their inventions to Mexico.[33] By allowing industrialists to bring their machines and technologies, without having to fear that their intellectual property rights would be compromised, political authorities anticipated that foreigners would set up large industries and hire Mexican citizens, all of which would transform the country's valuable raw materials into export-oriented goods. Howard Conkling's 1883 account of a visit to Guanajuato represents just one example of a growing trend toward embracing technology from abroad. Commenting on an American steam-powered mill set up in Guanajuato, Conkling asserted that the introduction of American companies and a Protestant work ethic would undoubtedly lead Mexicans to adopt similar approaches to improve their own economy.[34]

By advertising the ease with which one could register a patent and offering incentives to foreign investors, officials successfully increased the number of patents submitted to the Mexican office. For the 1901 world's fair in Buffalo, New York, Mexican officials published English-language pamphlets for investors as a way to promote the favorable business environment and patent laws.[35] The national patent office granted patents to numerous foreign inventors who had created fire safety devices: Berlin engineer Clemens Graaff patented his water storage tanks and gas-pressured hoses; Barcelona resident Domingo Biosea Galcerán submitted his design for handheld extinguishers; and Oakland, California, inventor Frank Vanoy Carman improved how the valves of fire hydrants opened.[36] From 1890 to 1896 the Secretaría de Fomento processed 1,044 patent requests. The nationality of the inventors broke down as follows: 307 from Mexico, 433 from the United States, and 304 from other countries, making more than two-thirds of the patents from foreigners.[37]

Sizing up Mexico's technological progress went beyond informal or anecdotal comparisons. In 1911, for instance, cartographer and geographer Mark Jefferson designed a formula based on international patents that sought to determine the rank of various countries' relative inventiveness. Out of thirty-five top invention-producing countries, led by Switzerland and Sweden, Mexico ranked number twenty-nine, ahead of Japan, Portugal, Turkey, Russia, and India. Even though this assessment ignored the contributions from countries that did not participate in international patenting processes, and most likely double-counted inventions patented in multiple countries, Jefferson's exercise nonetheless shows an estimate of Mexico's global rank in terms of technological developments in 1911.[38]

One engineer, astonished by how many patents came from foreigners, developed a scheme to increase domestic inventions. He suggested a three-pronged approach to stimulate innovation in Mexico. First, government officials needed to offer citizens access to credit. Second, national funds should be employed to develop a strong educational system, in major cities and small villages alike. Third, officials needed to provide incentives for local inventors and entrepreneurs to take risks and lead the charge for an improved technological existence.[39] None of those suggestions took root while Díaz was in office.

INVENTOR-ENTREPRENEURS

Even though local inventions never outpaced the imported machines and technologies from the United States and Europe, many inventors answered the call to develop a national tradition of innovation. Patents transformed inventing from the primarily anonymous effort it had been for centuries into an occupation in which intellectual creativity was publicly recognized.[40] Inventors Thomas Edison, the Wright brothers, and Guglielmo Marconi, who dazzled audiences with their displays of electric lights, expositions of human flight, and broadcasts of spoken sound, became emblems of a time when individual ingenuity could lead to social prestige.[41] As the fame of these inventors grew, people started to experiment and build in the hopes of gaining honor and glory.[42] In the last third of the nineteenth century, the majority of patents came from amateur entrepreneurs and not from scientists or engineers, meaning that a successful inventor did not need to flaunt a university degree to earn admiration for his designs.[43] Nevertheless, registering a patent did not necessarily mean success.

Most inventions at the turn of the twentieth century came from well-off businessmen from Mexico and abroad who had the capital to purchase supplies, and more importantly, who did not rely solely on the profits from their machines or devices to make a living. Usually, it was through their daily occupations that they stumbled on an area in need of improvement or streamlining.[44] Two industrialists, Díaz and Sala, thought that their experiences owning several manufacturing workshops in the capital made them different from most inventors, because their personal experiences created a keen awareness of the types of fires that tended to erupt in workshops and stores. As a way to share their knowledge with others, Díaz and Sala invented "The Flame Killer" (Mata Flama), a handheld extinguisher to put out factory fires quickly.[45] Similarly, M. Cisneros, a trained contractor, witnessed great flaws in the construction of roofs, and he used his occupational experience to develop a new process to build nonflammable roofs.[46] Wife, mother, and homemaker Carmen Chávez, who had a vested interest in the safety of her home, mixed together an assortment of household substances in order to produce a fireproof paste to paint onto rooftops and walls.[47] In each of these cases, the art of invention took on personal dimensions.

In Mexico City, the role of inventor-entrepreneur can be seen most visibly in Federico Gamboa's novel *Santa*, in its portrayal of the inventor Bruno Ripoll. In awe of his ability to read massive books full of technical jargon, tenants at the Gupuzcoana boardinghouse where he lives respect Ripoll for his academic achievements and his devotion to his craft. His fellow boarders envision the great social benefits that could come from his scientific musings and mechanical designs and give him space and quiet to conduct his work. The owner of the boarding house, Doña Nicasia, even extends Ripoll months upon months of free room and board because she trusts that his work will eventually earn him enough money to pay her back.[48] The boarders see in Ripoll what Mexico City residents saw in other engineer-inventors of the time: an educated, erudite man willing to use his knowledge to improve public life in the capital city. In many ways he typifies the late-nineteenth-century heroic inventor motif. The boarders adore him not because he has invented anything that has gained widespread acceptance but because of the possibility that he might do so. Much like the real-life inventors of his time, though, his promises fall short, and he cannot translate his knowledge into a practical result. His lofty plans to build a submarine never materialize beyond the small model he creates, and the boarders eventually lose interest in Ripoll and the owner kicks him out of her boarding house for not paying his rent.

The life of Daniel Blumenkron offers a factual counterpart to the fictional Ripoll and showcases how inventors used their talents to benefit themselves and their cities. Blumenkron, a native of the United States and permanent resident of the city of Puebla, was a distinguished member of society, a former US consul to Puebla, and owner of La Fortuna, one of the most profitable match warehouses in the country. He epitomized the ideal inventor-entrepreneur by combining scientific pursuit with profit motivation and by showcasing his work at international and national expositions.[49] Blumenkron used chemistry to combat environmental and safety problems that made matches either impossible to light or spontaneously combustible. Some physical settings, such as the tropics of coastal Veracruz, led to excessive moisture in the workrooms, making it impossible to light the match. Other environments, such as the deserts of Sonora and Chihuahua, desiccated factories, causing phosphorous matches to ignite spontaneously. With a thin layer of Blumenkron's patented varnish applied to the phosphorous tip of the match, no longer did matches cease to function in climates with temperature, humidity, and precipitation extremes.

Blumenkron followed all of the appropriate steps to improve the profits of La Fortuna while also gaining widespread praise for his innovations. He submitted his first patent application in 1878 but continued to fix its flaws and resubmitted in 1879.[50] In 1881, he put his invention into the US patent records, giving his name, invention, and country of residence international recognition.[51] Receiving much acclaim, he continued to promote his innovations in match making at national expositions such as the 1881 Exposition for the Society of Productive Workers of Jalisco in Guadalajara (Exposición por la Sociedad de las clases productoras de Jalisco) and the 1883 Toluca exposition.[52] Eventually, Blumenkron won awards and received international patents and recognition for his safety matches.

In 1880 his workshop, La Aurora, caught fire. Two boys who had been roughhousing and throwing stones at each other outside of Blumenkron's property inadvertently started it. When one boy threw his stone too hard, it smashed through a window in the business and landed directly on top of a large quantity of matches that had been drying after being dipped in phosphorous. The pile of matches immediately caught fire. Fortunately, Blumenkron kept a fire pump on hand for such emergencies, allowing him to suffocate the flames rather quickly.[53] Despite the fact that Blumenkron had built his success on inventing fireproof, safe matches, not even his business could be completely protected from this hazard.

As the ideal inventor-entrepreneur, Blumenkron saw in fire hazards destroying the city an opportunity to protect the population, earn a profit, and acquire prestige in the process. Blumenkron offers a case of a successful inventor. He was well respected in society and in 1880 he became an alderman in the Puebla Ayuntamiento; in 1893 he sat on Puebla's chamber of commerce (Cámara de comercio).[54] When Blumenkron died in 1895, journalists noted the extreme sadness that fell over Puebla.[55] Yet most inventors never gained this level of public recognition nor did they amass substantial profits for their creations. Despite getting the patenting office to recognize their inventions as novel, innovative, or necessary for national progress, the existence of their inventions was rarely even known by the wider public. Without being able to produce large quantities of fire safety technologies at affordable prices, few entrepreneurial inventors made money off of their inventions.

The role of individual inventors in society began to wane as inventing became assimilated into the research and development departments of large corporations. In the early twentieth century professional teams of engineers and scientists displaced the individual inventors who sought to profit from their ideas and earn public recognition. Major firms and corporations started to employ teams of scientists in their research and development laboratories to invent new technologies, helping to protect their company's monopolies on particular innovations.[56] By 1910, the famous cigarette company Buen Tono bought all the Mexican patents for cigarette-rolling machines, effectively monopolizing the process. No longer could individuals easily invent new rolling machines because Buen Tono had bought the rights and had a team of experts working to ensure their monopoly would hold. Similar monopolies developed in the areas of electrical lighting and chemical industries.[57] But even though large research and development groups eventually overshadowed individual inventors, in the last third of the nineteenth century inventors remained active in Mexico City and tried to profit from deep fears of fire hazards.

INVENTIVENESS: PATRIMONY AND PROFITS

From 1880 to 1910, the number of registered patents in Mexico skyrocketed, from 24 to 1,308.[58] Despite this dramatic increase, only one in four of the total patents during this period came from a native-born Mexican. That ratio was higher for fire safety patents, however. Between 1858 and 1914, roughly half of the fifty fire safety patents that arrived at Mexico's patenting office came from Mexican inventors. In their applications, domestic and foreign inventors spoke in patriotic terms of technology paving the way to progress and modernity. In effect, inventors reiterated the rhetoric of Comtean positivism adopted by officials, scientists, educators, and political advisers who believed that the application of science in public policy led to social progress. The historical context of a nation trying to achieve modernity by improving its industrial output and applying science to social problems shaped how technology developed in the capital.[59]

Fire control or prevention devices had the potential to benefit everyone in Mexico City, and this prompted local and national officials to encourage residents to invent machines, chemicals, or apparatuses to make the city safer. In select cases, such as the 1884 official request for the development of safety matches, political authorities looked to the public to solve problems through technology.[60] Government officials, businessmen, and investors identified match production in the capital as an area in dire need of redesign. In 1884, Porfirio Parra, a medical doctor, supporter of científico intellectuals, and advocate of positivist thought in general, asked the capital's citizens to submit designs for safety matches. Parra advocated the study of nature and the human condition as a way to learn how to predict things that could stifle production.[61] Match production offers a superb example of something that needed to be improved in order to conform to the demands of urban life.

Noticing increasing concerns about public safety, inventors tried to craft protection devices and promote them as works of national patrimony. The available matches, known for being highly flammable and spontaneously combustible, even presented severe health risks to workers in match workrooms. The first match, created by a British inventor in 1805, was marketed as the "Instantaneous Light Box."[62] Their ease of use made matches an everyday accessory in homes and businesses, but they could easily ignite. Knowing the government's aspiration to create a thriving manufacturing industry in the capital, one inventor remarked how safer factory settings, partially the result of his safety matches, would undoubtedly lead to more profitable national industries.[63] Other inventors commented on the steep prices that safety devices carried, and in turn they tried to use inexpensive materials so that everyone in the capital could afford to buy matches. They claimed that making safety more affordable to members of all social classes benefited everyone in the country.[64] Moreover, inventors tried to promote their ideas as emblematic of national progress, often naming their devices after political figures, such as Adolfo Martínez Uristas's fire extinguisher, the Extinguidor Universal Porfirio Díaz.[65]

Fire safety inventions offer insight into how people linked technological innovation with ideas about hygiene and progress. Mexico City physicians and sanitation inspectors asserted that public health and well-being offered the clearest example of progress in cities, especially during the last two decades of the nineteenth century, when foreign hygienists awarded the capital the dubious honor of being the most unsanitary city in the world.[66] Embracing the recommendations of physicians and miasma theorists about public hygiene, inventors designed healthy ways to prevent fires. Redesigning matches in attempts to make them less harmful and malodorous consumed the time and energy of many inventors throughout the country. Patent applications, matchbox labels, and newspaper advertisements praised safety matches that protected health,[67] while match vendors claimed that they only sold healthy matches, ones that blessed customers with the gift of hygiene.[68] Inventor E. M. Arzac asserted that he never used toxins to coat his matches, making them completely harmless to users' health.[69] Many inventors fully embraced the latest ideas about hygiene—or at least the rhetoric of hygiene—when patenting and promoting their inventions.

Following the lead of local governments that first initiated many urban reforms such as sanitation, garbage disposal, and waterworks as a way to protect the health of the population,[70] individual inventors also sought to protect the general well-being with technological innovations. Until the advent of the safety match, matchsticks had been coated with yellow or white phosphorus, a dangerous chemical that posed a serious risk to workers in match workshops and to anyone who lit matches on a regular basis. Employees at match workshops often developed phossy jaw (the necrosis and decay of the jawbone), which caused severe pain, irreversibly deformed the faces of its victims, and made it difficult, if not impossible, to chew food. This disease afflicted workers throughout the world.[71] After extensive toxicology investigations surrounding phosphorous poisoning in match factories, the US Bureau of Labor passed the White Phosphorous Match Act of 1913, which prohibited the manufacture and sale of matches coated with the harmful chemical.[72] Long before the implementation of phosphorus regulations in the United States, Mexican inventors understood the need to change match production practices for health benefits. Francisco Tadeo Linder, a Spanish-born owner of a match workshop in Mexico City, in his patent application, promoted his new formula used to coat matches as the healthy and safe alternative to ordinary matches.[73] Citing the prevalence of phossy jaw and the miasmatic fears that inhaling noisome odors caused disease, Linder and others accepted the prevailing but erroneous theories of airborne contagion of a number of diseases.[74] Concerned citizens, worried about the dangers of breathing offensive odors, encouraged inventors to produce matches with more pleasant smells, which were seen as inherently less dangerous.

Urban spatial factors also determined the types of inventions people deemed necessary for the capital. Announcing fires by ringing the nearest church bell caused problems in a large city. First, it delayed response time because an onlooker had to run to the nearest church, ring the bell, and hope that enough people heard the call for help. Second, bells rang nonstop, despite efforts to regulate them, to celebrate saints' days, to start mass, and to announce the coming of a religious procession. The message from the church bells often became blurred and knowing whether a bell rang to announce a church activity or in response to a fire meant that some residents inadvertently ignored desperate calls for help. This slow and highly inefficient method eventually was replaced in 1901, when the fire station received its first telephone, allowing firemen to hurry to the scene of the fire even if the fire occurred on the other side of the capital.[75] Yet again this meant that someone had to witness the fire and run to the nearest telephone (in 1901 it was most likely easier to find a church than a telephone). Unlike the noisy and ineffective public ringing of church bells, the use of the telephone to call the fire department ensured that only the professional fire brigade received information about the burning building.

To alert the public of an impending disaster more quickly, inventors created appliances that immediately warned people that a fire had begun. Some automatic alarm systems, like the one invented by Francisco E. Oviedo and then improved upon after his death by his wife, Matilde Rábago Viuda de Oviedo, alerted the household of a fire in the first moments that it ignited, making it easier to smother flames before they spread to adjacent rooms and buildings.[76] Other alarm systems went a step further by including automatic sprinklers. The first of these locally invented apparatuses came from Miguel Laimón, an electrical engineer from the capital. He constructed a thin metal box full of liquid carbonic acid that could be installed in any room regardless of the square footage of the space. The flames heated the metal box and sent a current from the heated metal to the copper wiring, which released a valve, thus dropping the liquid onto the flames.[77]

Mexico City electrician Guillermo Taverner improved upon the Laimón model by adding an alarm, but his newer system could only be installed in buildings with indoor plumbing, effectively limiting its consumer appeal.[78] The Taverner Automatic Extinguisher and Alarm contained a series of tubes with holes that could shoot out water in all directions to smother flames. When the temperature in a room rose to a particular point, mercury crept up the glass wall of the thermometer and triggered an electrical current to make a bell ring furiously. Once the bell started to ring, the electrical current continued along the wire and opened the valves that connected the building's water pipes to the tubes in the sprinkler system. Water from public waterworks rushed into the tubes of the Taverner automatic extinguisher and sprayed out through the holes and onto the flames (figure 5.3). Yet many problems remained with this method: the water needed to have pressure behind it, which required either a rooftop tank that used gravity to give the water natural pressure or an external steam- or electric-powered pump.[79] In several befuddling cases, the new machines used to extinguish fires actually started a fire. In factories in Saint Louis, Chicago, and Montreal, the rooftop tanks that provided sprinklers with water collapsed through the roof, causing their metal supports to clash together, spark, and start a fire.[80] Despite some problems with access to water, these inventions tried to confront the problems of living in a growing city where a house could become completely consumed by flames before professional help arrived.

Despite the many Mexican-made inventions for automatic fire alarms, General Félix Díaz, the nephew of the President and the inspector general in charge of the police and fire departments, complained to a US reporter that the fire stations remained severely underequipped, lacking an alarm system to alert the men to fires that erupted across the city.[81] This statement indicates that the invention of a technology did not always correspond to its adoption and use. Nonetheless, inven-

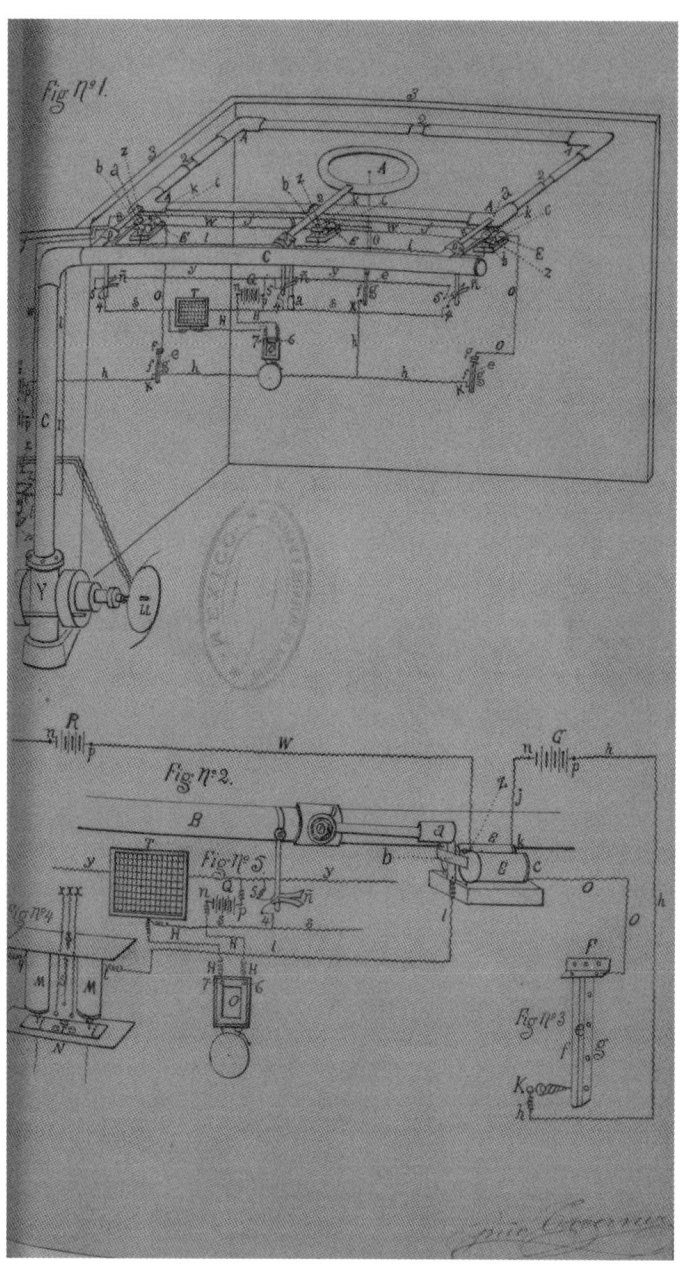

FIGURE 5.3. This drawing represents the complete installation of the Taverner automatic sprinkler system in a house or apartment. Guillermo Taverner, Alarma y auto-extinguidor de incendios "Taverner," September 1, 1909, AGN: Patentes y Marcas, leg. 308, exp. 26, fs. 8.

tors' efforts to create specific technologies show the particular areas that inventors thought needed improvement.

One of those areas of improvement was in theaters. During the first decade of the twentieth century, when cinema theaters began to experience horrific fires due to the advent of overheated projection rooms, inventors tried to profit from a growing hazard. Entrepreneurs made protection a business, seeing theaters as opportunities to profit from serious fire risks. English inventor Samuel West promised "magnificent and invaluable" results from his so-called innovation that included applying asphalt to rooftops and ceilings to prevent fires. West claimed that his asphalt application had saved from ruin a number of theaters throughout the United States, and was pleased to apply the asphalt onto the roof of the Teatro de Relox in Mexico City.[82] Mexico City businessman Carlos Villegas designed a machine to extinguish flames instantly when nitrate film caught fire. At the first sign of fire in the projector room, an employee could pull on a cord attached to a holding tank that would drop water directly onto the flames (figure 5.4).[83] Electrician Ricardo Rojas built upon Villegas's original innovation, adding a heat-triggered automatic sprinkler connected to lights within the theater intended to alert audience members to the fire.[84] Small-time inventors, in their efforts to earn a profit, thus pinpointed some of the major problems that plagued urban life.

While petroleum drilling had not become a problem for urban residents, many inventors used their experiences with urban fires to tackle petroleum-related hazards, like the 1908 fire in Dos Bocas, Veracruz. The increasingly profitable petroleum-drilling industry came at a high cost, and inventors saw an opportunity to make the industry safer while lining their pockets with profits from this emerging enterprise. Inventors tried to protect against the risk of physical injuries and property loss by creating fire safety instruments designed to extinguish flames in petroleum wells. Mexico City industrialists Reinaldo Rodríguez Arce and Everardo Rodríguez Arce developed a device to extinguish flames that had started inside wells, thus saving valuable petroleum and preventing devastating fires. Their invention required the use of long, extendable metal tubes connected to a large storage tank of water. Functioning much like a home fire sprinkler, in this system, once a fire erupted, workers would extend the tubes into the flaming well and flip a switch that released the water onto the flames.[85] Similarly, Guadalajara resident and engineer Ernesto Fuchs also addressed the risks of coming into contact with burning wells of petroleum and

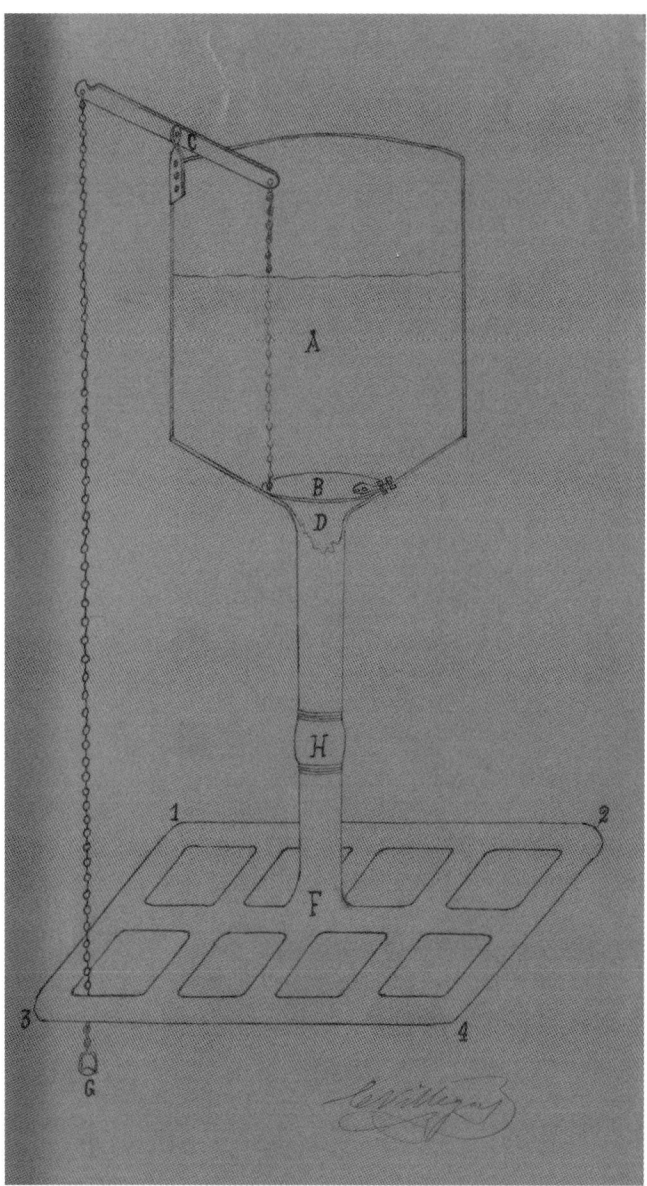

FIGURE 5.4. Carlos Villegas's invention to put out fires in the projection booths of movie theaters was intended to be mounted on the ceiling over the projector. If the film caught fire, the projectionist would pull the chain and release the water directly on top of the projector. Un aparato para sofocar incendios en las casetas de los cinematógrafos, December 6, 1907, AGN: Patentes y Marcas, leg. 308, exp. 18, fs. 3.

INVENTING PROTECTION

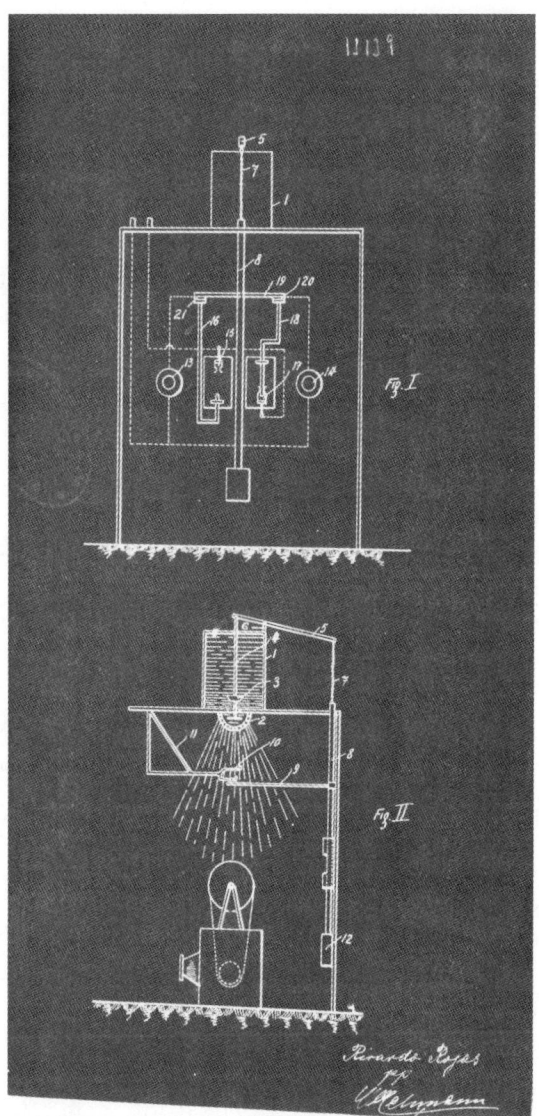

FIGURE 5.5. Ricardo Rojas's automatic sprinkler system included a piece of cotton wrapped around the poles marked 9 and 10. After a fire erupted, the flames would burn the piece of cotton and release the two poles, causing a reaction that would release the water in the tank. Additionally, when the poles dropped, a weight trigger attached to a light switch turned on the lights in the theater to alert the audience. Un aparato automático contra incendio de películas cinematográficas, June 29, 1912, AGN: Patentes y Marcas, leg. 308, exp. 35, fs. 4.

designed a gun that shot fire grenades full of baking soda and water into the well.[86] Both of these inventions employed technology to keep people out of harm's way while also attempting to preserve valuable reserves of petroleum.

THE USE OF INVENTIONS

It is important not to conflate the history of technology with the history of invention, because the former tends to emphasize inventions that worked and were successful. Studies of innovation and inventors often overlook how people used the invention, or whether or not anyone ever adopted it for daily use.[87] Few documents exist to help to answer the questions about the adoption of technology. Municipal reports provide some clues about the adoption of firefighting technologies. For example, city council members authorized the purchase of fire grenades and extinguishers for public buildings. In official fire reports, the fire chief noted the number of fire engines, hoses, pumps, and extinguishers that firemen employed to put out conflagrations.[88] Engineering reports also hint at how business owners adopted technology; for instance, in their reports city inspectors listed the type and number of extinguishers located in each public theater.[89] Even health inspectors spoke of the fire pumps, hoses, and handheld extinguishers they used to wash the streets and reduce the spread of germs during a typhus epidemic.[90] While the majority of these accounts do not indicate the particular brand or origin of the firefighting equipment, they nonetheless demonstrate that these technologies became common accessories in the city, functioning as viable fire control methods for the residents.

Many patented inventions never gained widespread use, but the study of patent applications offers a way to assess the technological demands of society.[91] When analyzed side by side, patent applications can show the small alterations made to improve a product.[92] Some inventors addressed high production costs by using more affordable parts that could be found within the country. Others marketed their extinguishers as less cumbersome than previous models. Inventors always had to consider the bottom line and think about ways to meet consumer demand. The small changes to existing devices demonstrate how social needs influenced technology.[93]

Consider fire extinguishers. During the last decades of the nineteenth century, handheld extinguishers evolved from glass balls filled with water and thrown onto flames to intricate machines with levers, tubes, and chemical reactions that shot water long distances. In 1884, after numerous experiments, Juan Septien, a professor of chemistry and pharmacy at the National University, submitted a patent for his fire-extinguishing device. The chemist, best known for being an official in the Ministry of Development (Secretaría de Fomento) and for creating popular antisyphilis syrups,[94] boasted that he had invented a method to free society from the disastrous consequences caused by large-scale fires. Septien presented his design of a thin-walled metal ball filled with water, sulfuric acid, and baking soda. The professor promised that once the safety pin had been pulled out and the grenade had been thrown directly upon flames, it would release an effervescent liquid that would put out fires of any size or intensity. President Manuel González eagerly approved Septien's patent request for a ten-year period.[95]

With only slight differences to the design, another inventor was able to hold a simultaneous patent for a device almost identical to Septien's fire grenade. In 1884, the Wadsworth, Martinez, and Longman Company of New York patented in Mexico its glass fire grenade, which did not require the user to pull out a safety pin. Rather, one needed only to throw the grenade on the fire hard enough for the glass to break, allowing the flames to be exposed to the baking soda mixture.[96] Both fire grenades could suffocate a small fire within a matter of minutes because the presence of baking soda effectively deprived the fire of oxygen, making the devices far more effective than water alone. In addition, the effervescence also spread the water over a larger surface area, making the substance suitable for fire extinguishment. Wadsworth, Martínez, and Longman's advertisements claimed that half a dozen grenades worked better than six to eight gallons of water.[97] The Ayuntamiento, aware of the grenades' small size and ease of use for those with no prior operating knowledge, ordered ten boxes from the New York–based company.[98] This represents one of many instances in which Mexican officials encouraged its citizens to invent technologies as a way to contribute to national pride yet continued to purchase devices from abroad. Understandably, the Ayuntamiento's decision to buy the New York grenades could have been due to the fact that Juan Septien never had the production capacity to fill an order for ten boxes of grenades.

While businessmen claimed that fire grenades put out huge fires, evidence proves otherwise. Philadelphia surveyor and fire underwriter Charles John Hexamer vehemently opposed fire grenades because it

FIGURE 5.6. The cover image from the Wadsworth, Martínez, and Longman catalogue showing how to use the handheld grenade. Unas granadas de mano para extinguir incendios, September 2, 1884. AGN: Patentes y Marcas, caja 25, fs. 9.

was difficult to break the glass and release the liquid. He joked that it would be more efficient to first locate a corkscrew to remove the cork from the bottle than to attempt to break the grenade as the product's inventors suggested.[99] Even the Harden Hand Grenade's own advertisements tempered expectations, warning, "Do not expect too much of one Grenade. Use enough at the beginning to do the work promptly and completely. They are so cheap that you can afford to use them unsparingly."[100] Fire grenades remained popular among residents through the 1930s, but today, antique collectors who consider the technology a novelty are the main consumers of fire grenades.

Complaints that the fire grenades extinguished only the smallest fires encouraged inventors from around the world to design machines that held larger amounts of water while remaining portable enough for one man to carry. Newspapers advertisements raved about the benefits of handheld fire extinguishers, such as the "Extinguidor de incendios de Babcock," which, according to company propaganda, had single-handedly put out more than 1,200 active fires.[101] Pamphlets that the

FIGURE 5.7. Mexico City fireman with a handheld fire extinguisher.
Overland Monthly, 117.

Babcock Fire Extinguisher Company distributed worldwide included dozens of testimonies about how well it worked. Joseph L. Perley, chief engineer of the New York Fire Department, wrote that the Babcock extinguishers were "beyond a doubt, the best apparatus ever invented for self-protection, and if they were in general use, many millions annually would be saved from destruction by fire." R. A. Williams, the fire marshal of Chicago, echoed that sentiment, noting, "Every person should be their own Fireman, and arm themselves with them." The company gained official recognition for its invention at the Cincinnati Industrial Exposition of 1874, where the Babcock Fire Extinguisher Company won prizes for the best fire engine and best fire extinguisher.[102] Even underwriters and insurance agents praised the device and urged their customers to purchase the extinguishers for their homes and businesses.[103]

The earliest extinguisher models had large tanks of water with secondary, smaller tanks of gas or carbonic acid inside. The Minimax brand extinguisher, which the Puebla Ayuntamiento purchased to protect its public buildings,[104] represented this type of device, and customers often complained that the tank of gas took up too much space, which, they thought, could have been filled with more water (figure 5.8). Addressing these concerns, W. Graaff & Compagnie of Berlin developed a new model in 1905 and redesigned it in 1906. A metal cone full of water held a small glass tube full of baking soda and acid; with a swift push to the external plunger, a sharp metal rod broke the glass tube, causing the baking soda, sulfuric acid, and water to mix and create a chemical reaction that produced gaseous bubbles and propelled water out of the cone and onto the flames (figure 5.9).[105] Such modifications to existing devices showcase how inventors sought to improve existing designs.

Frustrations with everyday technology led some consumers to design better methods, specifically for use in the capital city.[106] Businessman Ruben Martí reworked the Graaff & Compagnie extinguisher and created one that held more water and could be built out of easily replaceable parts. Instead of a glass tube that needed to be imported from Germany, Martí employed two locally made glass bottles that he filled with baking soda. Martí proposed that rather than puncturing the glass and thus making it unusable for the future, a lead ball be put inside the bottle as a stopper. As soon as the operator turned the bottle right side up the lead ball fell to the bottom, allowing the water, baking soda, and sulfuric acid to mix and bubble (figure 5.10). Know-

FIGURE 5.8. "Minimax. The King of fire extinguishers. The extinguisher that never fails. Adopted and recommended by all the monarchs, governments, and fire departments." *El Siglo Diez y Nueve*, December 26, 1874, 4.

ing the difficulties in obtaining imported parts quickly, Martí asserted that extinguishers needed to be inexpensive and accessible, and by offering easily interchangeable parts he encouraged people to buy his invention.[107] S. F. Hayward and Company from Philadelphia also incorporated the lead-stopper method in its Rescue Fire Extinguisher. In this case, the operator needed to invert the extinguisher to dislodge the stopper, allowing the sulfuric acid to mix "with the soda solution and the gas formed creates a pressure sufficient to throw a stream of the chemical 50 feet."[108] Many inventors used chemistry to improve existing designs, often employing the same chemicals that had produced so

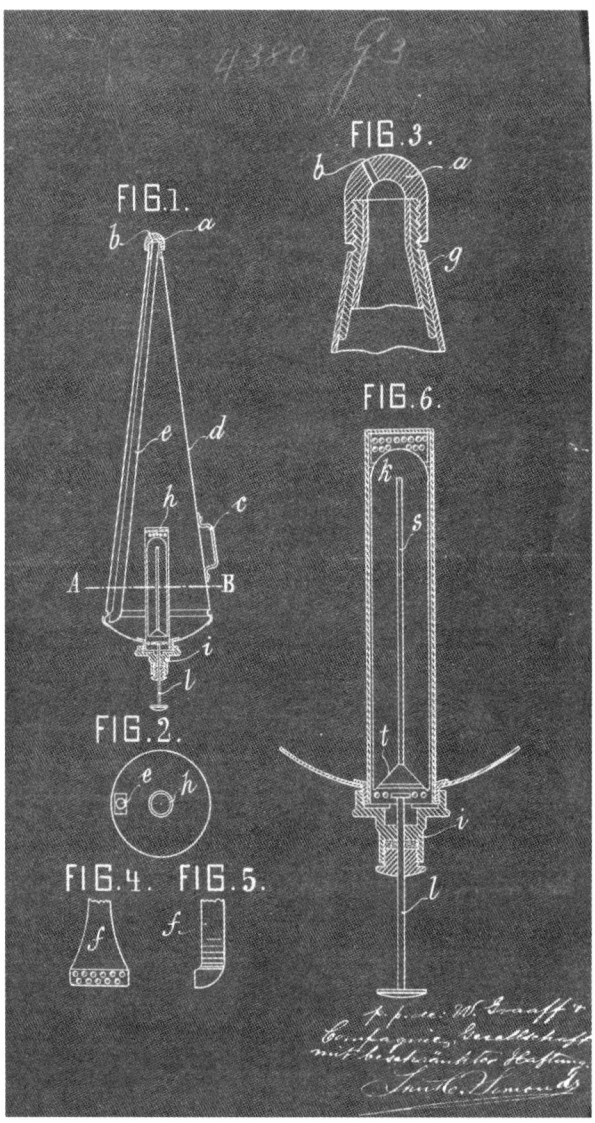

FIGURE 5.9. Drawings that accompanied W. Graaff & Compagnie's patent request showing the components of their handheld fire extinguisher. Figure 6 is a drawing of the glass tube filled with carbonic acid that can be punctured by the metal rod. W. Graaff & Compagnie, Gesellschaft mit beschränkter Haftung, ciertas mejoras en extinguidores de incendio, February 21, 1905, AGN: Patentes y Marcas, leg. 308, exp. 11, fs. 5.

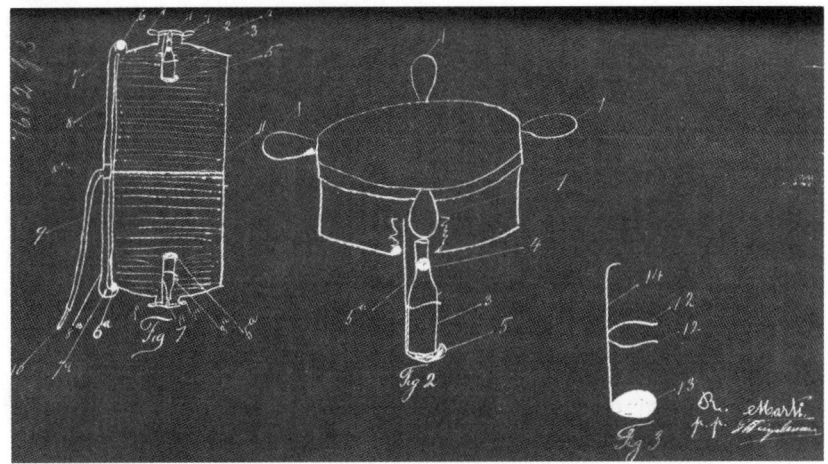

FIGURE 5.10. Ruben Martí's extinguisher with easily replaceable glass soda bottles to hold the carbonic acid. Un extinguidor para incendios, January 21, 1908, AGN: Patentes y Marcas, leg. 308, exp. 19, fs. 4.

many fires in the first place. Here science and chemistry played a role in both generating fires and extinguishing them.

As the devices became more efficient in smothering flames, they also became more difficult to use and required some operational knowledge and specific maintenance procedures. Extinguishers with lead stoppers required that operators take on the dangerous task of re-filling the tank with sulfuric acid, which often resulted in skin burns from splashing acid. W. S. Darley and Co. of Chicago skirted this prob-lem by creating compressed-air pumps, but these pumps required more operator knowledge and higher maintenance standards.[109] The Bab-cock Extinguisher Company wrote its clients and detailed the main-tenance procedures, which included discarding the old water from the apparatus and "recharging" it with a new vial of carbonic acid. They even reminded clients that the extinguishers were "machines" and that "legs and brains they have not, and therefore cannot look after them-selves."[110] A small problem with the Graaff & Compagnie's 1905 extin-guisher, for instance, was that it could not be stored on its side because it would leak. The hoses in Ruben Martí's invention easily got clogged with the debris in water. Some inventors claimed that Domingo Biosea Galcerán's extinguisher and Díaz y Sala's Flame Killer had small capil-lary holes that slowly released gas and left the extinguishers without enough water pressure to put out fires.[111] Making a firefighting device consistently reliable was a difficult task and often required that users actively maintain the machine and replace parts when necessary.

Inventions did not always work as expected, a phenomenon that remains difficult to detect in patent records. In their attempts to convince the patent officials to authorize their machines or products, inventors rarely spoke openly about imperfections. Nevertheless, instances of inventions not performing as expected suggest that the lofty promises of health, youth, safety, or wealth often did not materialize. Nicolás Zúñiga y Miranda, an inventor and opposition presidential candidate who ran against Porfirio Díaz in 1896, 1900, 1904, and 1910, designed several machines to predict earthquakes. His lack of success in predicting natural disasters made him one of the capital's most recognizable cranks and the butt of many jokes among both popular and elite classes.[112] Moreover, popular stories of other inventors accidentally killing themselves while experimenting with combustible chemicals perpetuated ideas that not all inventions worked, and that some were downright dangerous.[113]

Sometimes technology did not work because machines, ideas, and know-how did not transfer well from one place to another. The best example of a blundered attempt at installing technology in Mexico City occurred at the newly established Peralvillo slaughterhouse, where the owners installed an assembly line model identical to the one they had in Saint Louis. Immediately, things went awry. The conveyer belt system dropped carcasses on top of workers, equipment jammed, and drains did not work properly, leaving several inches of animal blood covering the factory floor. Over a dozen workers died the first year and the owners blamed poor factory replication abilities, inconsistent Mexican and American engineering training, and the country's lack of compatible screws, nuts, bolts, and other tools to repair the machinery.[114]

Such technological failures often occurred as a result of environmental or social barriers. The National Iron and Steel Works eventually closed its mill because the country lacked sufficient supplies of wood, coal, and ore.[115] The operating capabilities of some technologies worked well, yet they still failed to be adopted by society due to high costs or lack of skilled workers to operate machinery. Whether or not a gadget worked and people bought it, the fact that more and more people spent the time inventing fire extinguishers, sprinklers, and alarms indicates that fire hazards presented a problem in the capital and that people confronted them through technological innovation.[116]

Despite these homegrown attempts at fire safety innovation, most Mexican citizens chose to adopt devices from foreign companies. One way to see this phenomenon is through Sanborn fire maps. These maps offer a window into the private lives of residents by opening up buildings and peeking inside to see what safety precautions they purchased for their protection. The maps help to answer the question about whether or not technologies were adopted and used in the city. While alarms and sprinklers did not appear in many homes, some larger businesses, such as La Concordia Clothing Company, did invest in these technologies.[117] Surprisingly, businesses purchased an array of extinguishers from different companies. While owners from La Tabacalera and El Buen Tono cigarette factories purchased Worthington pumps from New York, owners of the Zolly Hermano's hat manufacturing workshop preferred the Triplex brand power pump from Chicago to guard the felting machine. In regards to handheld chemical extinguishers, La Victoria Woolen Mill used Miller Chemical extinguishers of Chicago, the Mexican Gas and Electric Light Company imported Babcock handheld extinguishers from Chicago, and La Gran Unión Distillery favored the New Peoria Chemical Fire Extinguishers from Peoria, Illinois. In 1905 only one company, the Iron Works and Foundry, still used glass hand grenades. Limiting their inventories of extinguishers and sprinklers to the biggest factories, the Sanborn fire maps show that these business owners preferred to import safety devices rather than purchase one of the many Mexican-made products.

A budding culture of inventiveness in Mexico promised inventor-entrepreneurs that they could gain acclaim by tinkering with machines and devising scientific experiments. Their inventions could be showcased at local, regional, and international expositions, where inventors could be awarded prizes and recognition for their achievements. Despite an overall increase in the number of Mexican-made inventions that received patents, few capital residents ever purchased or used these machines and technologies. Whether or not a given invention worked well or was ever purchased, the volume of inventions, measured through patent requests, government documents, and newspapers, suggests there was a growing need for fire safety in the capital. Moreover, the rise of the inventor-entrepreneur also signaled the growing desire to purchase safety precautions on an individual level. Nowhere does this trend become more evident than in the realm of fire insurance.

CHAPTER SIX
INSURING PROGRESS

It [insurance] is the handmaid of commerce and the guardian of industry.

National Board of Fire Underwriters, *Fire Insurance*

A popular anecdote retold among insurance agents begins with a stranger asking an insurance adjuster what he does for a living. The insurance adjuster quips that he is in the business of "buying ashes."[1] This humorous characterization of insurance has some truth to it, but more importantly, it highlights one of the underlying concerns that clients had when purchasing a policy—that they were not buying a tangible product, only an agreement that the leftover ashes from a burned structure could eventually be transformed into a new home or business. Chicago underwriter J. A. Fowler characterized the insurance industry a little differently, saying that he sold a product that "frees the mind from a burden of care and anxiety as to a specific danger."[2] It was the freedom not to worry about future risk, and instead to worry about more pressing daily concerns. Those who sold this type of protection waxed rhapsodic about their chosen profession: "Insurance may justly be deemed one of the noblest creations of human genius. From a lofty height it surveys and protects the commerce of the world. It scans the heavens; it consults the seasons; it interrogates the ocean, and, regardless of its terrors or caprice, defines its perils and circumscribes its storms. It extends its cares to every part of the habitable globe, studies the usage of every nation, explores every coast, and sounds every harbor."[3] By selling peace of mind and a promise of protection against unforeseen accidents in the future, insurance agents gave clients the confidence to invest and the freedom to take business risks. Without insurance, one underwriter wrote, "commerce would be paralyzed," because it is the security upon which business enterprises rest.[4] In a moment of industrial growth and development, insurance agents, business owners, and government officials alike hailed the benefits of fire insurance because it facilitated capitalism in the modern, liberal period.

Liberalism gives primacy to the individual. The writings of Comte and Spencer emphasized self-improvement, personal achievement, and self-reliance, which have become the underpinnings of liberal thought. Ideas about private property, individual rights, and economic freedom became appealing in comparison to collectivist approaches that were often characterized as unreliable and unpredictable. In the case of fire, prior to the liberal era, community-based protection that spread the economic risk among families and villagers was the standard protocol. Yet the assumption that everyone dutifully practiced fire safety in homes or workplaces often had disastrous consequences—even more so when uncertainty and risk grew with new manufacturing techniques and fuel options in cities.[5] The ethic of private responsibility meant that the fate of one's property rested firmly in the hands of the individual, thus making community responsibility an afterthought.

In the search for reliable and calculable fire safety, something that seemed the opposite of community-based fire protection, individuals found what they were looking for in fire insurance. Throughout much of nineteenth-century Mexico, *socorros mutuos* (mutual aid societies or friendly societies) provided the only financial assistance in the case of fire, accident, or death. This collective approach relied on the voluntary support and cooperation of group members. Coworkers in factories or neighbors contributed money to a fund or offered services and goods in kind to help ailing members in times of need.[6] By 1844, various mutual aid societies had 318 capital workers as members. Over the course of several decades, they began to merge with cooperatives, which had a different approach to managing funds that included investing to generate growth, rather than saving and redistributing when necessary.[7] Increased hazards in the city made it harder for mutual aid societies and cooperatives to take care of all of their members' needs, yet they were not completely swept away during the Porfiriato. The late-nineteenth-century context, which blended the environmental factor of greater fire risks and the ideological factor of a drive for individualism, offered private insurance companies a timely opening to privatize safety and profit from risk. Unfortunately, when that safety had to be purchased and was no longer provided by a community, the poorest members of the city became even more vulnerable to fire hazards.[8]

THE SCIENCE OF RISK

By purchasing insurance, businessmen and homeowners attempted to predict and prepare for emergencies. In doing so, they took responsibility into their own hands and refused to leave anything up to chance.[9] While the products that insurance companies sold did little to eradicate urban fires, they did offer policyholders a sense of relief that they had fulfilled their individual responsibility to protect property and profits. It was not just the insurance policy that alleviated clients' fears of fire loss; it was the way in which companies determined rates and premiums. To those unaware of how insurers determined rates, insurance resembled gambling more than science, but as one underwriter explained, "On the contrary, its animus or intent is the very antithesis of gambling. In gambling we seek the danger and excitement of uncertainty; in insurance we seek the repose and safety of certainty."[10] Insurance actuaries calculated and quantified something as seemingly unquantifiable as risk. Mathematical equations, statistical formulas, and financial theories helped actuaries determine "the science of casualty,"[11] whereby they could deduce the likelihood of a fire. Even though schools for fire inspectors did not exist until several decades into the twentieth century, many inspectors had prior technical experience in factories or warehouses that gave them the kind of knowledge that was crucial for the world of insurance.[12] To uncover the physical causes of fire, underwriters and inspectors equipped themselves with handbooks that categorized various workshops, machinery, and chemicals based on their hazardous potential.[13] Some handbooks recommend not insuring particular businesses, including ones that made and sold candles, soap, candy, and fireworks. Other businesses had to meet requirements before an insurance policy could be issued, including the installation of fireproofing construction and automatic sprinklers in theaters.[14] These establishments had a high propensity for starting fires and historically had not been profitable for insurers.[15]

One thing that actuaries could not quantify easily was human error, or what insurers call "moral hazard." A moral hazard, as opposed to the aforementioned physical hazard, is obscure and unquantifiable, and it includes everything from carelessness to malice to pyromania.[16] It also includes a rather common psychological problem that insurance companies both past and present face: the tendency of people to become careless and neglectful of fire safety precautions once they have purchased insurance. Many policyholders misunderstood insurance as protection from fire, leading some to imagine that as soon as they signed the policy their property was safe from fire.[17] In their minds, after purchasing an insurance policy, the burden of protecting oneself from fire shifted from the home or business owner to the insurance company. Insurers found "moral hazards" so troubling because they were difficult to predict and yet so prevalent, leading one actuary to claim, "A very great cause of fires is the wicked recklessness of our people."[18] In 1867 a British fire assessor, Henry Daniel, commented before Parliament that between 50 and 70 percent of fires are not accidental, meaning that the majority of insurance claims are fraudulent.[19] If these estimates were correct, fires actually became more dangerous as insurance became more common, because property owners handed over control and responsibility to the insurance policy and neglected basic duties that would prevent fires.

Calculating individual morality was difficult. To determine moral fortitude, an insurer interviewed neighbors and researched business records to assess the potential client's character and financial solvency. While the insurance agent had the primary responsibility of deciding whether or not to insure a client, companies advised their agents to deny insurance to "bad or dishonest persons," "strangers," or "threatened persons." Insurance companies suspected that anyone with a questionable background was more likely to start a fire intentionally, or if said person had an enemy of ill repute, the enemy might seek revenge through incendiarism. Unfortunately, individuals living in structures adjacent to disreputable people also became uninsurable, since fire did not respect structural boundaries and could quickly leap from building to building. A messy or unorganized establishment may also have led an insurance agent to become concerned with moral hazards, because a dirty or unkempt establishment reflected the owner's character and therefore suggested a higher risk of catching fire.[20] An insurance agent's judgment, although not calculated or precise, played a large role in determining one's suitability for insurance and the rates a policyholder would be expected to pay.

PRIVATE FIRE INSURANCE IN MEXICO

Initially, the fifteenth-century enterprise of insurance emerged to protect maritime business from the high incidences of shipwrecks and piracy.[21] With the success of maritime insurance, businessmen in growing European cities demanded similar types of protection from financial loss due to fires.[22] While fire insurance emerged in the early eighteenth century, it was not until the latter part of that century that purchasing fire insurance became more or less a standard procedure for most industries in major European cities.[23] The Industrial Revolution helped account for this dramatic increase, as more and more factory and business owners understood that there was a greater risk of fire with new fuels and machines. Thus, they purchased insurance policies to protect their investments. By the 1860s, discussions of insurance in Mexico claimed that "the most cultured nations" of the world had already established a robust insurance industry, and that Mexico was distressingly behind.[24]

Insurance addressed the financial problems posed by environmental hazards, which spurred government officials to align with private insurance companies in order to facilitate abundant capitalist growth. Having the ability to insure a business against misfortune gave some entrepreneurs the impetus to invest and expand their businesses. By the time Emperor Maximilian arrived in Mexico in 1864 his advisers warned him that local industry was completely unprotected, often citing the lack of fire and life insurance as one of the principal reasons for the faltering economy. The insurance industry had become a standard entity in Europe and the United States, yet Mexico's insurance industry was limited to maritime protection. Predicting that insurance could encourage the rise of business and commerce, Maximilian allowed the first life and fire insurance companies to enter the Mexican market in 1865. Don Florentino Romero established La Bienhechora (The Benefactor) life insurance company and La Previsora (Foresight) fire insurance company with headquarters in Mexico City and offices in twenty-three cities across the country.[25] In addition, the companies promised that insurance would "moralize the masses, inspiring in them ideas of order and precaution."[26]

Government officials thought Mexico was ripe for this type of behavioral change and protection from the unforeseen. The Ministry of the Interior (Ministerio de Gobernación) also granted Romero a license to publish a weekly periodical, called *La Mutualidad*, which extolled to readers the importance of insurance to improve one's material condition.[27] Shortly after the success of La Bienhechora and La Previsora, two more Mexican companies arrived in the capital: La Mexicana (fire insurance) and El Porvenir (life insurance), both directed by Don Joaquin Acebo.[28] All of these insurance companies were of Mexican origin and had at least fifty thousand pesos in cash reserves to cover claims.[29]

Over the next several decades, with the establishment of more foreign businesses in Mexico, a steady stream of foreign insurance companies arrived to provide protection. In 1865, the Home Colonial Fire Insurance of London set up an office in Mexico City and catered to the growing population of foreigners. Liverpool and London, the Manchester Fire Insurance Company, La Nacional Prusiana, Sociedad Hamburgo-Bremense, and Phoenix Assurance Company made inroads into the growing Mexican market for fire insurance. A 1903 English-language directory of businesses in Mexico City listed twenty-four fire insurance companies, all of which were foreign owned.[30] The presence of foreign insurance companies showed foreign investors that Mexico provided a safe investment climate, and that insurance companies were there to hedge investors' risks. The Díaz administration saw in fire insurance a way to strengthen and transform the country's economy. The flourishing export industries of cotton, sugar, tobacco, and henequen, along with a textile industry that consisted of more than 140 manufacturing workshops, the mining industries of silver and gold, and the investments made by foreigners in areas of railroad construction and petroleum excavation, all required protection from fire risk.

Not every home or business owner qualified for insurance. Each company clearly articulated its coverage guidelines and what it would and would not cover. The National Commercial Code (Código Comercial) guided insurance protocol and demanded that every item a policyholder wanted insured had to be documented and appraised. All titles, merchant documents, paper money, precious metals, artwork, furniture, and almost anything that had monetary value could potentially be insured.[31] Typical fire insurance policies of the time covered damage caused by gas explosions, faulty electrical wiring, and most accidental fires, whereas all intentional fires and arson fell outside the insurance companies' responsibility and became a legal matter to be decided in municipal courts. The concept of "correlated risk," still used in today's insurance industry, also applied to nineteenth-century Mexico. Correlated risk refers to numerous dangers occurring simultaneously, such as an earthquake that then causes fire or flood damage, as was the case in the 1906 San Francisco earthquake fire, or manmade dangers such as war. Nineteenth-century Mexico's reputation for political instability and chaos meant that insurers had serious concerns about correlated risk. Every insurance company operating in the country stipulated that policies became null and void during civil wars, foreign invasions, military force, or popular riots. In addition, most policies did not cover damage and losses from natural disasters, including volcanic eruptions, hurricanes, floods, earthquakes, or forest fires.[32] In the case of both war and natural disasters, entire communities could be razed, leaving an insurance company unable to fulfill all the insurance claims at once. Without these stipulations, foreign insurance companies never would have come to Mexico because the risk would have been too great.

In order to court investors and advertise Mexico's safe investment climate, the Díaz administration outlined and instituted a national insurance law. Intended as a way to protect the interests of the people who bought the insurance, the Law Concerning Insurance Companies attempted to guarantee that insurance companies would not mislead their clients with promises of security that the companies could not actually fulfill. According to the 1892 law, the Ministry of Finance (Secretaría de Hacienda) had to approve all paperwork regarding the creation of new insurance companies. This provided a way to control an industry that provided no tangible products, just the promise of security. To do this, the secretary of finance demanded copies of all the insurance policies sold in Mexico, the names of all the employees, and banking information regarding the company's assets showing that the company could cover the costs of major insured accidents.[33] If the company failed to produce these documents, it would be fined between fifty and five hundred pesos.[34] The Ministry of Finance monitored each insurance agency closely and even hired investigators to enforce the Law Concerning Insurance Companies with on-site inspections every six months. Due to the growing demand for insurance and insurance regulation, the Ministry of Finance created the Department of Insurance to expand its reach and continue to protect against insurance fraud and abusive insurance companies.[35] Not surprisingly, insurance companies preferred to manage their businesses without this level of government intervention.

At the end of the nineteenth century, foreign companies dominated the Mexican insurance market. By 1897, fourteen of the fifteen fire insurance companies operating in Mexico were foreign-owned. Together, all fifteen established the first Mexican Association of Fire Insurance Agents (Asociación Mexicana de Agentes de Seguros contra Incendio) in 1897 as a way to voice their concerns collectively and to protect their interests. Local authorities quickly began to notice a startling trend regarding the operations of foreign-owned insurance companies in Mexico. Nearly every time a foreign-owned business insured by a foreign company burned to the ground, the owner of the defunct business took the insurance money and left Mexico. This meant that the owner reinvested funds elsewhere and left Mexico with little more than a pile of burned rubble.[36] Domestic insurance companies tried to buck this trend and subsequently encouraged economic nationalism. For example, La Previsora's guidelines stipulated that all claimed insurance money had to be used to rebuild a home or business in Mexico, which ensured a cycle of reinvestment in the Mexican economy and its future.[37]

INSURANCE AND SOCIAL CHANGE

Insurance agents, much like inventor-entrepreneurs, claimed that their products could elevate national development and benefit all social classes.[38] As La Fraternal, a free monthly bulletin about life and accident insurance in Mexico claimed, "A nation that does not produce cannot be rich nor have commerce and trade, and conversely, nations that produce more have greater opportunities to foment human activity and intelligence."[39] One way of facilitating production was through insurance, or what La Fraternal called the "noble, humanitarian, and patriotic" act of providing citizens with the promise of financial safety.[40] The bulletin also published stories about people who died or experienced crippling accidents but fortunately had insurance to save their families from financial ruin. In its first issue, the editor explained that the objective of the publication was to show that accidents happen all over the republic, to anyone, irrespective of race, class, or ethnicity.[41] A workplace injury, unexpected death, or home fire could destroy a family's savings. By telling heart-wrenching stories of loss, La Fraternal tried to persuade readers to buy insurance to prevent hardship. According to the publication, widows and orphans, abandoned with no insurance policy, suffered the most and became financial drains on society.[42] Insurance advertisements reminded everyone that socially burdensome widows and orphans could have easily been protected had their families taken precautions and bought insurance.[43]

Attacks on individual responsibility became especially vehement after a fire in 1901 at the Mercado del Volador, in which most of the merchants completely lost all the goods in their stalls, leaving them in abject poverty. While La Fraternal applauded the capital's public charities for generously donating money and food to victims, it simultaneously condemned the market vendors for not carrying any type of insurance to protect themselves from misfortune. After describing how ill prepared the market vendors were for such a disaster, the editor of *La Fraternal* went on to describe how in Mexico City, "the rich as well as the middle classes are sufficiently cautious and do not ignore events that may ruin their interests," suggesting that the poorest people in society willingly left their fate to chance.[44] In a market that had a history of devastating fires that ruined people's lives, the vendors were once again ill prepared and blamed for their misfortune.

The way *La Fraternal* framed the outcome of the 1901 Volador fire as a product of individual negligence highlights how some interest groups in society, specifically those making money off of fire protection, wanted to transition fire safety away from collective responsibility and toward individual responsibility. Thus, it formed part of a larger process in the liberal era in which individuals bore responsibility for their protection and relied on private capitalist firms, rather than community networks or public programs, to provide disaster relief. Here, the privatization of safety refers to both the development of private capitalist firms (insurance companies) and the evolution of an ethos of private responsibility. For Mexico, there were benefits and shortcomings to this kind of logic. On the one hand, under Porfirio Díaz and other liberal reformers Mexico owed a lot of its growth and development to this pattern of capitalist development. Investing in Mexico City became less risky. On the other, people who could not afford to purchase fire safety (like the Volador merchants) had fewer and fewer options and became more vulnerable to urban risk.

Insurance agents not only touted their products as effective ways to avoid financial ruin, they also claimed that insurance would increase human ambition and ingenuity. *La Fraternal* went so far as to say that, without insurance's promise of safety, a society could never excel.[45] Starting a new business came with a number of unforeseen risks that dissuaded people from striking out on their own and insurance companies offered a way to alleviate the fears about one of those risks: the chance that fire might destroy their buildings, products, or machinery.[46] León Rass, the owner of a stocking workshop in Puebla, certainly felt compelled to use insurance as a safety net. By 1911 he had purchased four fire insurance policies from different companies in order to cover all of his assets in the case of a fire. In a 1911 court case that documented the insurance investigations of a fire in his stocking workshop, Rass explained that each time he wanted to expand his operation he purchased additional insurance to protect his investments.[47]

Despite his eventual success sorting out his insurance claims and getting all four companies to agree to cover his losses, Rass, like many other insured persons in Mexico, had to participate in a lengthy investigation. In accordance with the Commercial Code, investigators hired by the civil court assessed the validity of each insurance claim.[48] When the Puebla stocking business burned to the ground, the four insurance companies collectively hired one detective to determine that an accidental fire, not an intentional one, had occurred and to inventory the remains to confirm that Rass had not lied about the assets he lost. Three investigators, one hired by the civil court, another by the insurance companies, and one by Rass, rummaged through the charred debris of his workshop and wrote down everything that could be salvaged. They scrutinized the owner for counting too many of one type of stocking and not enough of other types of stockings. They also made detailed maps of the building, inventoried everything that was not burned, and sent their findings to Mexico's civil court and their respective insurance companies. Rass eventually received his insurance money, albeit several hundred pesos less than he had originally estimated.[49]

The case of a Mexico City sweets shop that burned in 1895 highlights the propensity of insurance companies to avoid paying insurance claims. On October 23, 1895, Juan María Colomic's sweets shop became engulfed in flames at 1:10 p.m. The workers were all on their lunch breaks when the three-story building caught fire and neighbors reported that the first sign of fire was a great flash of light that escaped through three windows in the back of the workshop. Soon thereafter the flames stretched ten meters above the building as they ate through the wooden roof. The flames extended to two adjacent homes, forcing the occupants to flee in terror. Two hours after the fire began firemen were able to calm the blaze and isolate it to the bottom floor of the sweets workshop, where it eventually died down after destroying almost everything in the shop.[50] Juan María Colomic had preemptively purchased two ten-thousand-peso fire insurance policies from North British and Mercantile as well as Liverpool and London and Globe to give himself the relief from worrying about one risk of starting a business. Collecting that purchased protection proved more difficult than the insurers had advertised.

Agents from the insurers fought in Mexican courts for several years to absolve themselves from having to pay Colomic's insurance claim. The agents scoured the commercial code and hunted for reasons to avoid paying the claim. When writing a detailed letter to Mexico's Supreme Court, the agents laid out their reasons for not paying the Colomic claim, arguing that the policy stipulated that the companies only had to pay a claim in the case of a "total fire," whereas the Colomic fire was a "partial fire."[51] Three expert mechanics examined the site of the fire and determined that, even though the machines used to make the sweets were still standing, they were completely unusable and only suitable as scrap metal. They concluded that it would cost more to repair the machines than to replace them. Architects also reviewed the site and examined the existing walls, and determined that indeed the fire was total, not partial, as the agents had tried to claim. At the end of the Supreme Court's investigation the judges ruled that the insurance companies not only had to pay Colomic but also had to pay the Supreme Court for the fees and costs associated with the investigation.[52]

The insurance agents did not stop there. According to Colomic's policy, he was obligated to report the fire within fourteen days, yet according to article 781 of Mexico City's Commercial code, Colomic needed to file his claim within twenty days of the fire. The Commercial Code also stipulated that the twenty days did not include Sundays or national holidays. The insurance agents argued that the insurance policies' filing regulations should trump the Commercial Code's regulations because it was a signed agreement between Colomic and the companies. Additionally, the insurance agents found the holiday exception unreasonable because Mexico celebrates so many holidays, and they exaggerated that if Mexican holidays were taken into consideration, Colomic would not have needed to file the claim for several months, at which time the site of the fire would have been contaminated and the investigation useless. The Supreme Court also ruled in favor of Colomic because he reported within twenty days (excluding holidays and Sundays) and an investigation had been done immediately after the fire. According to the ruling, Mexico City police officers and other city employees conducted a thorough investigation and inventoried all of the machines and materials that survived the fire.[53] The insurance companies had little recourse and finally paid Colomic on May 11, 1898, almost three years after the fire.[54]

Based on the initial price of insurance, along with the likelihood of a costly and lengthy process to receive payment on a claim, it is clear that not everyone could afford to purchase this home or business security. When the Mexico City cigar manufacturing plant La Rosa del Valle suffered an extensive fire, it took the Royal Insurance Company more than a year to settle the claim. During that year, the Fornaguera

Ortíz brothers, who owned the plant, could not operate at the same level of production that they had prior to the fire, undoubtedly losing money during the insurance investigation.[55] After the corner store Los Heróicos Boeras burned on the night of February 18, 1902, Liverpool and London refused to pay the owner, listed as Señora Romano, the three thousand pesos for her losses. Romano prepared to fight the decision and appeared at her court hearing accompanied by her husband. Her lawyer recounted Romano spouting off articles 451 and 551 of the Código de Procedimientos Civiles, which protected her rights as a policyholder. After three months, and with the help of her family, Romano was able to have a reopening party at her corner store, yet still awaited the insurance money from Liverpool and London. Six months after the fire, the court forced Liverpool and London to pay her, stating that the Commercial Code could not be any clearer about this matter.[56] A certain level of financial stability and individual initiative to research lengthy commercial codes and legislation must have already existed for insurance to be a viable option. Plus, on many occasions, government intervention was necessary to force insurance companies to deliver the promises of individualized, purchased protection.

The privatization of protection put a price tag on urban safety. Seizing on the fact that Mexico had a market of eager individuals looking for protection, insurance companies knew that they could profit from the misery caused by catastrophes. Insurance companies started to promote their products to concerned citizens and foreign investors, advertising their products in newspapers, giveaway calendars, and monthly newsletters.[57] These forms of publicity allowed insurance agents to increase consumer demand for insurance through compelling stories of loss. They aroused fears about future danger and reaped the rewards from an uneasy population. The question of how to quantify risk also led to the development of another form of disaster expertise—insurance mapmaking—that helped define spatial boundaries of safety and risk.

MAPPING SAFETY

Insurance offers a way to assign ideas about risk and safety onto city plans. The United States Sanborn Fire Insurance Map Company, in 1905, conducted an extensive cartographic review of Mexico City's central business district to track vulnerability to hazards. Aware of the growing market for fire insurance, the Sanborn Company sent survey-

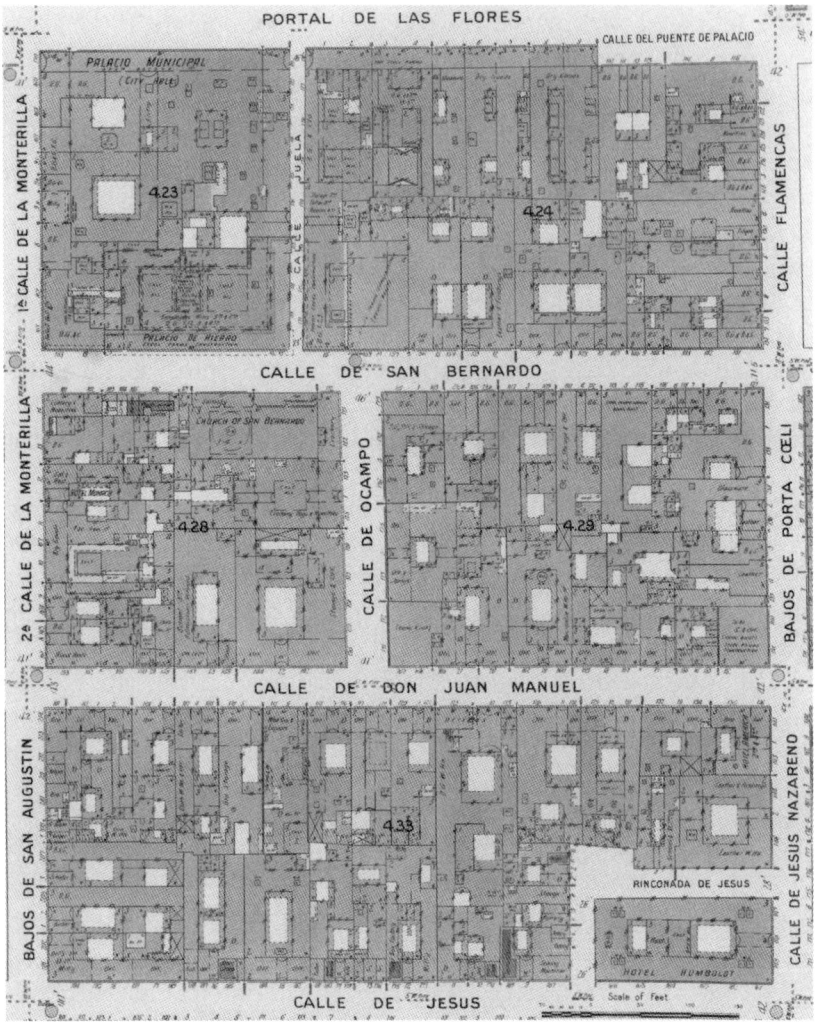

FIGURE 6.1. "City of Mexico, Plan 9," *Sanborn Fire Insurance Maps*, 1905, Perry-Castañeda Library Map Collection. Courtesy of the University of Texas Libraries, University of Texas at Austin.

ors to the capital to map the city, intending to sell the maps to insurance actuaries who would analyze the data to calculate premiums for clients. Actuaries and mapmakers transformed natural disasters into formulas and measurements that could be calculated and predicted.[58] Determining the likelihood that a building would catch fire required that Sanborn surveyors spend months measuring buildings, plotting fire hydrants, and drawing color-coded plans of the city. Mapmakers scrutinized and cataloged every component of the built and natural environment. They drew every boiler, stairwell, chimney, and door located in the business district. They also noted the physical characteristics of the buildings, whether they were made of brick, shingles, wood, adobe, or steel. In addition, they mapped things that could not be seen from the street level: underground pipes, rooftop cisterns, household taps, and handheld fire extinguishers.[59]

While similar mapping projects had become common practice in North American cities and towns with populations of more than one thousand people, Mexico represented a new market for the Sanborn Company. In the first decades of the twentieth century, mapmakers branched out into Mexico City (1905), Ciudad Juarez (1893, 1898, 1900, 1902, and 1905), Ciudad Porfirio Díaz (today Piedras Negras) (1905), and Mexicali (1921). In most US and Canadian towns, the Sanborn Company returned to update its maps every five to ten years, always using standardized cartographic markings and colors in order to show change over time accurately.[60] Unfortunately, the Sanborn Company documented Mexico City only once, in 1905, limiting this source of information on urban transformation. Nevertheless, these maps offer a visual representation of how fire regulations and technologies became woven into the fabric of the city.

Government, residents, and insurance companies worked in tandem to prevent and prepare for future disasters. Earlier fire codes noted a growing interest in using wooden construction and therefore restricted these building practices. The Sanborn maps indicate that, at least in the business district, the fire code worked because almost all the buildings were made of "a natural stone, the common material for walls is 'tepetite' composed of fragments of pumice loosely bound together by a clayey substance."[61] Here the mapmakers presumably meant *tepetate*, a naturally occurring material used for buildings and common throughout the Valley of Mexico.[62] Mapmakers noted that asphalt covered all of the streets in the city district, while stone covered the outlying districts, suggesting that the wooden boardwalk-style sidewalks that had been common only a few years before had become a thing of the past. For the most part, by 1905 major business owners followed the strict fire codes that limited the presence of combustible fuels in the city center. Businesses with large volumes of flammable substances or machinery capable of producing deadly explosions, such as the Mexican Gas and Electric Light Company, La Victoria Woolen

Mill, La Tabacalera Mexicana cigarette factory, and La Gran Unión Distillery, sat outside of the designated fire zone. The fact that residents followed these protocols suggests that the citizenry felt it had a collective obligation to prevent fires and reduce fire risks in the city.

Often, modern amenities such as gas lighting and electrical apparatuses caused the greatest harm, forcing businesses that used those materials to make drastic alterations to how they operated on a daily basis. La Indianilla Compañía Limitada de Tranvías Eléctricos de México, which functioned as the main electrical generating station for all the streetcars and shops in the city, functioned like a little city inside the metropolis.[63] Equipped with tenement housing and wooden shanties that sat next to the city's drainage canal, as well as pulquerías and corner stores, La Indianilla employees made it a priority to protect the space where they worked and lived. Due to the large number of electrical generators, air compressors, and large boilers that could easily explode and ignite the fuels in nearby coal sheds and petroleum tanks, the company took extra precautions to prevent fire hazards. The managers of La Indianilla organized a company-run fire brigade to control fires. The brigade, made up of La Indianilla employees, had its own fire engine and maintained the company's sixteen fire hydrants, six wheeled chemical extinguishers, and thirty one-gallon hand extinguishers. To train plant employees how to maintain composure and prevent chaos in the case of a fire emergency, La Indianilla's fire department conducted fire drills three times per week. La Indianilla functioned as a microcosm of the city, and the insurance maps indicate that the owners and managers mimicked many of the policies that Mexico City officials had implemented for the entire city.

In addition to showing how residents responded to municipal legislation, the Sanborn maps visibly display the engineers' craft, especially public works projects that involved water. Each map indicates the diameter and location of the city's underground pipes and whether they were attached to fire hydrants. The surveyors even went into individual homes and small businesses to inventory the city's household water supply. The maps recorded all of the cisterns in the city, including information about how many gallons of water each one held. Besides infrastructural improvements, the maps indicate that the city engineers' concerns about theater safety translated into real change. Even though the Teatro Principal housed a large wooden stage, theater owners listened to the inspectors' demands and protected the establishment from the hazards of fire. Except for the stage, the Teatro

Principal's auditorium, lobby, and dressing rooms were made of fire-resistant stone. Three hydrants, powered by electric pumps, sat upstage behind the curtain and could receive up to two thousand gallons of water from a rooftop cistern.[64] In contrast to the Teatro Principal's fire safety precautions, one of the secondary theaters, Teatro Arbeu, did not have nearly the same degree of fire protection. Arbeu's fire control consisted of one small water pump on the roof of the building and its location next door to a carpenter's workshop, a type of business especially known for catching fire, disturbed inspectors.[65]

Engineers worked to create a safer city, but their efforts did not reach all residents equally. Take the case of the Mercado del Volador, which Sanborn mapmakers referred to as the "Thieves' Market."[66] In their maps, surveyors provided commentary about the market, a structure with a long history of catching fire and wreaking havoc. The market housed more than one hundred permanent stalls and a large number of temporary, constantly moving booths that were poorly constructed, often made of cheap boards with canvas roofs. Despite the numerous Volador market fires that had occurred over the previous half century, the map shows that few safety improvements had been made to protect all the hat makers, fruit vendors, tobacco sellers, and candlemakers. Only four small hydrants protected the entrances to the market and one large cistern sat outside the market near its southwest corner.[67] For comparative purposes, the central post office of the capital, approximately one-quarter of the size of the Mercado del Volador, was flanked on all sides by twenty-six fire hydrants, each attached to a twenty-five-foot hose.[68] Presumably, far more people entered the Mercado del Volador than the central post office on a daily basis, yet because of the post office's importance as a government building and the heart of national communication, it received more protection. Regardless of the market's close proximity to the National Palace and the Metropolitan Cathedral, officials continued to overlook it, ignoring the safety concerns of the market's vendors and customers.

In addition to surveying dangerous workrooms and mills, Sanborn mapmakers provided commentary about general customs found inside family dwellings, thus offering a lens into the private lives of capital residents. Unlike many homes in the United States and Canada, Mexico City dwellings usually did not have chimneys.[69] Nor did residents use wood to cook their food and heat their homes. In some ways this was beneficial, since chimneys and burning wood had a tendency to cause serious fires. However, the Mexico City alternative may have been worse. The insurance maps note that capital residents tended to cook with charcoal in pit grills in courtyards for their daily needs. This common household fuel source produced less smoke than wood but also burned hotter and longer.

Despite the mapmakers' attention to details such as the types of fuels and stoves most commonly used in private residences, not everyone in the city wanted to divulge information to Sanborn representatives, and many business owners refused to allow the surveyors to enter their buildings. Based on the businesses that refused admission to surveyors we can assume that the owners feared receiving fines and penalties for breaching the city's fire code. Several furniture stores, trunk shops, photography studios, and the large Monte de Piedad National Pawn Shop refrained from being included in the city survey. Reluctance to disclose information about potentially harmful business practices threatened the health of the city, reflecting individual freedom to avoid surveillance that might interrupt business practices.

The presence of a foreign insurance map company entering the Mexican market offers another example of the role of privatization of safety. Insurance companies or businesses that had some vested interest in quantifying the levels of risk a place carried could purchase the Sanborn maps. In an era that valued scientific approaches to calculating everyday experiences like fire, the Sanborn fire maps were incredibly valuable. They also spoke to liberal sensibilities that valued individual responsibility and individual freedom, as well as another major corollary of liberalism—capitalism. The owners of La Indianilla took responsibility into their own hands and purchased technologies and equipment to protect their capital investments, whereas the owners of Monte de Piedad National Pawn Shop understood that individual freedom meant they did not have to abide by the requests of a private company.

Many praised insurance for facilitating capitalism and encouraging investment, which greatly benefited Mexico City's development near the turn of the century. A 1912 encyclopedia about insurance and fires claimed that "without [insurance] the wheels of progress would be badly blocked."[70] To extend this hyperbolic metaphor, if insurance greased the "wheels of progress," government intervention assured that the wheels stayed on track and delivered the goods that had been promised. Mexico's national government had to build an entire regulatory infrastructure to handle insurance and protect customers who purchased it. On numerous occasions, individuals required the intervention of government officials to ensure that promises of private insurance were fully realized. But government regulation did not necessarily harmonize with the private insurance industry's goals of individual self-reliance, or more specifically, individual self-preparedness for disaster. Nonetheless, citizens needed the regulatory oversight by public officials in order to fulfill market transactions. The multiple layers of fire protection required that the apparatus of government work in tandem with private interest.

CHAPTER SEVEN
HEALING THE HAZARDOUS CITY

Where would such a life end? In the hospital and the cemetery, of course, the last port of call in every life's voyage, whether virtuous or sinful.

FEDERICO GAMBOA, *SANTA*

The final layer of the story of fire safety in Mexico City goes down to the level of the human body. Damaged lungs, infected wounds, scarred tissue, psychological trauma, and excruciating death can all follow exposure to flame or smoke.[1] High injury and mortality rates associated with burns decreased the population's productivity and the country's vitality. During a period when a healthy population was emblematic of national progress and stability, treating and preventing illness and disease took on a new importance. The local environment, which had become exceedingly susceptible to fire, piqued the interests of members of the medical establishment who incorporated burn treatment into their curricula. This academic training reflected a paradigm of prevention that sought to contain and manage disease and illness and worked alongside similar initiatives driven by hygienists and public health officials.[2]

Scholars of Latin America have long addressed the relationship between urbanism and disease to explain the health risks of living in cities,[3] yet few have examined the public health effects of factors other than disease, such as occupational and industrial hazards.[4] In a rapidly industrializing city, many residents confronted harmful toxins or dangerous equipment on a daily basis. Increasingly, capital residents began to understand the changing dynamics of their city through experiences of pain and suffering on bodies. In his analysis of industrialization in Japan, Brett Walker sees a similar phenomenon and explains that workers who interacted with toxins were often plagued by bodily pain, and "certain kinds of pain serve as internal 'biological indicators' of a poisoned landscape."[5] Similarly, in Mexico City, residents were reminded of the city's hazardous nature through the memory of being burned and the permanent visual markers of scarred tissue.

By the 1860s Mexico City mirrored other modern cities throughout the world in housing several hospitals and a national medical school where students studied the newest European approaches to medicine. A community of experts emerged who had a common belief that attending universities, reading medical journals, and earning state-supported credentials made them the most equipped practitioners in the city to handle the health concerns of ailing residents. Not all health experts were included in this community. Local healers, for example, were deemed to lack the training and background to be part of the network of physicians and public health officials who made lasting decisions about how to treat illness and disease in the city. Regardless of the distinction between professional and lay, credentialed and uncredentialed, or modern and traditional, in most homes it was local knowledge about how to treat burns that prevailed. Wives and mothers tended to the majority of fire-related injuries, seeking out professional assistance only after home remedies had failed. Increasingly negative attitudes that everyday citizens held about hospitals and physicians led some patients to get help elsewhere. Physicians, healers, and burn victims, armed with different training and experiences with fire, tried to transmit their healing knowledge onto a city suffering from fire hazards.

TRAINED PHYSICIANS TREATING BURNS

Industrial accidents and burn cases had become not only more frequent but also more severe due to the abundance of flammable substances that could be found in homes and workplaces. Petroleum fires and gas explosions created deeper and more extensive burns, which prompted burn treatment to become a necessary skill that medical students practiced regularly. The Medical Sciences Association (Establecimiento de Ciencias Médicas), founded in 1833, eventually transformed into the National School of Medicine in 1854. The school taught students how to care for burns in order to prevent infection and death, and over the decades students were able to choose specializations ranging from bacteriology to dermatology to therapeutic medicine, all of which dealt extensively with burns.[6] The large faculty and access to specialized classes encouraged students to expand their concern beyond saving a burn victim's life to experimenting with improving the appearance of the burnt skin with ointments and skin grafts.[7]

The available records about burn cases are based on data from hospital registries or public health institutions and only include those

TABLE 7.1 Statistics showing the number of people admitted to Hospital Juárez and what type of injury or ailment led them to seek medical aid.

Injury or ailment	Admitted	Died	Percentage that died	Mortality rank
Blunt force trauma	1,013	205	20.24	1
Burns	406	72	17.73	2
Gunshot wounds	499	88	17.64	3
Simple fractures	930	104	11.18	4
Sharp object wounds (nails, needles, hooks)	4,576	423	9.24	5
Dislocations	229	11	4.80	6
Puncture wounds	2,927	138	4.71	7
Bruises	1,748	74	4.23	8
Contusion wounds	25,884	191	0.74	9
Stab wounds	7,693	54	0.70	10
Bite wounds	892	6	0.67	11

Source: This table comes from Federico García Sepúlveda's 1896 MD thesis, "Estadística general del Hospital Juárez, 1888–1895."

burned people who sought help from government-run agencies, disregarding those who tended to their wounds at home. Physicians noted that they often treated nearly dead burn victims who had waited several days to visit a medical professional, suggesting that the majority of burned people shunned professional medical treatment and sought attention in hospitals or physicians' offices only as a last resort, usually after an infection had taken over or as secondary symptoms arose. By overlooking the vast number of burns treated outside of professional hospitals or medical offices, the available statistics offer an incomplete picture. Nevertheless, some health officials attempted to record quantified data. For example, the records from 1860 to 1864 show that, of the 406 patients who were admitted to the Hospital Juárez for burns, seventy-two died.[8] In relation to other ailments, few people were admitted to the hospital for burns. Yet burns had the second-highest mortality rate, behind only blunt-force trauma and slightly higher than gunshot wounds, which shows how difficult it was to save a burned patient.

Burns were included in the introductory training of how to treat basic lesions, which also covered bruises, wounds, fractures, and dislocations.[9] Students mastered how to heal burns in several of their required courses and in their practicums in hospitals. While interning at hospitals, many students had to repair damage done to patients who had initially endured minor wounds but insisted on treating the wounds at home without the use of antiseptics.[10] Students prepared for such experiences by taking classes, the majority of which required that they read textbooks from Paris that detailed the newest and most effective ways to approach ailments and illnesses. In 1862, the amount of textbook space allotted to burn treatments spiked when Professor Juan N. Navarro adopted Auguste Nélaton's book *Élémens de pathologie chirurgicale*.[11] Beginning that year, the students in his *clínica externa* (external medicine) class learned how to classify burns based on color, smell, scars, and perceived pain. The text warned students to watch for rising fevers and shock, and to amputate fifth-degree burns on limbs in order to stave off infection and save the life of the patient.[12] The six degrees of burns that students memorized in the late nineteenth century roughly correspond to modern-day burn definitions, where first-degree burns are mild and resemble sunburns and fifth- and sixth-degree burns are so deep and lethal that they are tantamount to death and physicians usually diagnose them only after conducting autopsies.[13] The courses and the textbooks all stressed the importance of timely treatment, and municipal officials followed suit by ordering physicians to rush to the fire scenes to begin treating burn victims immediately.[14] Using antiseptics or cleaning wounds carefully by extracting any dirt, fragments of clothing, or other debris coincided with the modern practice of preventative medicine.[15]

Despite the growing emphasis on cleaning wounds to prevent infection, thorough cleaning and dressing did not always heal the patient. In his thesis on external medicine, Pedro Martínez Garza retold the stories of four men he treated for burns in the Hospital Juárez in 1895. Each patient was an adult male with a strong physical constitution who had endured first-, second-, and third-degree burns on his body. In each case, Martínez Garza washed the burned skin daily with an antiseptic solution, applied powdered bismuth, and wrapped the skin in cotton gauze. Administering a dry powder on top of the injury was a common healing practice that physicians used at the time. Bismuth and iodoform (chemical elements used in medications and cosmetics) were used interchangeably and rapidly dried the blood into a crust, which formed a protective layer to prevent infections.[16] Three of the patients' burns healed within a matter of weeks, leaving behind white scar tissue, pink spots of newly formed tissue, or folded and leathery skin. The fourth patient, despite receiving the same treatment as the other burn victims, died in the hospital after three days.[17] When students and physicians had followed all the protocols and taken the requisite measures to prevent infection but nonetheless saw their patients die, experts often diagnosed shock and mental duress as the cause of death.

Serendipity and coincidence often helped most in finding useful treatments. In 1838, Dr. Edward Greenhow of London treated a boy who had fallen into a pot of boiling pitch that severely burned the boy's arms. Only able to remove the pitch from the boy's hands, Dr. Greenhow remembered the work of British physician Dr. Thomas Baynton, who treated ulcers by spreading warm oils and plant materials directly on the wounds.[18] Dr. Greenhow decided to leave the pitch on the boy's arms and monitor the skin below. Three weeks after the accident, the boy's arms had no discharge or foul odor and the pitch had begun to flake off, leaving the arms completely healed and covered with new skin. By contrast, the areas of the boy's hands where Dr. Greenhow had previously removed the pitch had only just begun the healing process in the same period.[19] Dr. Manuel Andrade, a physician in Mexico City, used Baynton and Greenhow's initial findings and altered the treatment slightly, using natural materials like lime, which contains antibacterial properties, and potatoes, which he could find in abundance.[20] The organic material that Andrade used included lime juice and potato pulp mixed with oil, which he then brushed onto a burn five times a day to help ward off infection and repair the damaged tissue.[21] All of these oil-based treatments were precursors to the work of Dr. Barthe de Sandfort, a French physician who treated burned soldiers on the front lines in World War I with a mixture of wax, resin, and paraffin that he called ambrina. Earlier treatments used dry methods and powders, but Sandfort's wax approach kept moisture locked in, with the wax acting as a second skin that aided against infection, and this approach fundamentally changed how doctors treated burns.[22]

Mexican physicians had access to all the most recent medical journals and books published in Europe, making them well aware of the latest treatments for burned skin, but they made adjustments for local needs.[23] In the early 1870s, Swiss physician Jacques Louis Reverdin and British physician George David Pollock both published widely about their successes with skin grafting. They attached small pieces of healthy skin over burns or ulcers to repair the wounded flesh.[24] In response to these initial grafting experiments, physicians from around the world experimented with skin grafts to try to find the most successful methods.[25] After learning of these new medical developments, two Mexican physicians, Dr. Luís Muñoz and Dr. Carmona y Valle, immediately started to experiment with the skin-grafting methods of Reverdin and Pollock, adding slight alterations to the original methods. After cutting a small piece of skin from the body of an ailing patient, Carmona y Valle and Muñoz gently placed the healthy skin on top of the cleaned wound, pinning it in place with fine gold pins and wrapping everything with a thin sheet of sanitary plastic.[26] Yet that process did not yield any discernable results. Dr. José María Bandera, a physician at the Hospital San Andrés, also experimented with grafts. In a medical thesis from 1874, one of his students wrote about Bandera's approach to grafts, which was to first apply collodion (an adhesive syrup made of ether and alcohol), which acted as a glue to bind the graft to the skin.[27] The collodion approach, first attempted by French physicians and later modified by Dr. Bandera in Mexico, became especially helpful when using larger grafts or combining many small pieces of flesh to form a mosaic over the wound.[28] The globalization of medical knowledge led many of Mexico's most prominent physicians to experiment with grafting and adopt methods to better suit local needs.

Globally, all physicians had the same problem with skin grafts: the graft tended to die and decompose before any adjoining could occur. Attempts to keep the graft healthy and alive led some physicians to experiment with a type of skin grafting known as the pedicle approach, which left the skin partially attached to its donor. This method took partially detached skin, usually taken from a nearby healthy section of tissue, and stretched it over the injured skin, securing it with collodion, sutures, or wrappings. The pedicle approach kept the original piece of skin alive longer, allowing the blood vessels to continue to function and to feed both the healthy and the injured flesh. Sometimes physicians tried to use tissue from another source, either a living human or animal, thus attaching the burned victim to a living donor. Using a separate tissue source in the pedicled skin graft rarely worked because of the extreme difficulty in keeping the donor and the recipient immobile for several weeks. In 1869, French surgeon Charles-Emmanuel Sédillot cut a flap of skin from a hairless dog and sewed it onto the burnt hand of one of his patients. This experiment failed because the excessive movement on the part of the living dog would not allow the graft to take root. In 1880, Chicago physician Dr. E. W. Lee had initially tried to treat a ten-year-old girl's burns by attaching a flap of skin from her brother's thigh onto her burnt back, leaving the two siblings bound together for several days. After realizing that the sibling approach was not working, again due to excessive movement by the victim and donor, Dr. Lee then decided to try the pedicle method with animals. He immobilized two living lambs in casts of plaster of Paris and cut large flaps of skin from their backs, leaving the flaps still connected to the living animals. He attached the animal skin to the girl's back, but she died within a few days. Nevertheless, the autopsy found that the lambskin graft had been successful and sections of the animal skin had adhered securely to the girl's tissue.[29]

Because of the obvious problems with the pedicle method, more often than not physicians tried free skin grafts, which meant fully detaching the piece of tissue from the original donor before securing it onto the burn or wound. After having read Reverdin's 1869 speech to the Paris Surgical Society and Pollock's 1870 article about a skin transplant in London, Mexico City physician Dr. Luís Muñoz started to experiment with free skin grafts at the San Andrés hospital. In an article he wrote for *La Gaceta Médica*, the leading medical journal in Mexico, Muñoz related how only days after he had read about Pollock's findings in London, a patient who could benefit from a skin graft had been admitted to his hospital. On September 24, 1870, thirty-six-year-old Guadalupe Palma arrived at the Hospital San Andrés with a severe case of gangrene on her right arm, where deep-seated abscesses in the subcutaneous tissue ate away at her flesh. The brownish infection was serious, extensive, and painful. Knowing that the woman would likely die from her infection, Muñoz decided to try the new skin-grafting method he had read about. He cut five pieces of flesh, half the thickness of the dermis, from Palma's left thigh and applied them to various parts of the affected arm, securing them in place by wrapping the arm in plastic. Several days later four of the five transplanted pieces of skin had died. Within a few weeks, the gangrene had spread to Palma's internal organs and she died in the hospital. An autopsy of Palma's body revealed that the fifth piece of skin was securely attached to her wound and had healed nicely. Even though the experiment ended in the patient's death, Muñoz claimed success and concluded his article by saying that skin grafts would add "intrinsic value" to Mexican medical practices and would help save people from accidents, trauma, and infections.[30]

In spite of the interests that grafting had created in the medical world, physicians found it difficult to experiment with the new method because few people would willingly give up large chunks of their healthy tissue. This insufficient supply of grafting material led many physicians to seek viable skin in unlikely places. Swiss physician Jacques Louis Reverdin employed rabbit skin, while French physician Dr. Dubrueil used pigskin and the abdominal walls of dogs to test the grafting method.[31] In 1871 US physician Thomas Bryant successfully grafted the skin of a black man onto a white patient, which caused significant controversy at the time, with critics equating it to using an animal's skin.[32] In 1874 the head of the external medicine clinic at the Hospital Juárez suggested that physicians should start experimenting with animal grafts.[33] The use of animals to cure disease had already become essential to Mexico's antirabies campaign when, in 1888, Eduardo Licéaga traveled to the Pasteur Institute in Paris to learn how to make the rabies vaccination from infected rabbit brains.[34] For the case of burns, most Mexican physicians preferred instead to pay people for skin samples rather than using animal tissue. Dr. Muñoz tried to offer his patients monetary compensation for chunks of their skin, yet most of them refused. In order to continue his grafting experiments, Muñoz resorted to finding skin donors in need of financial assistance, presumably looking toward vagrants and beggars to assist him.[35] Licéaga, like many Porfirian experts at the time, was full of contradictions in his approaches to health, hygiene, and medicine. While in 1874 he had been scouring the streets looking for homeless people to offer up their skin to his scientific experiments, by 1891, after he became president of the SCC, he framed healthcare as a humane act.[36] The Porfirian drive to modernize, whether through engineering, city zoning, or health, was full of incongruities and contradictions.

CALMING PATIENTS: PHYSICIANS TREAT TRAUMA

Without patients, doctors could not study illnesses and learn how to treat pain.[37] The words, descriptions, and reactions of sufferers could tell physicians how to treat illness.[38] Valuable patient-physician interactions have shaped how physicians have understood and treated expressions of physical, sensory, and even emotional pain. Physicians have to treat burn patients in a series of steps, and each step along the way requires that the doctor and patient build a relationship. In the case of burns, physicians were first concerned with preventing infection to save the patient's life, then promoting mental stability, and last improving the appearance of the burnt skin. Firsthand experiences with burns represent the broader problems of life in a city where emblems of modernity and progress—new power sources, machinery, and fossil fuels—could have a negative effect on the population's health and safety.[39]

To make an educated diagnosis, doctors scrutinized the social lives and statuses of patients in order to understand the origin of certain illnesses and the best course of treatment. By bringing the patient's daily existence into consideration, doctors often adjusted treatment plans according to their patient's race or class.[40] Despite evidence of such inconsistencies, doctors denied that they applied varying criteria in evaluating and treating patients. Saving and improving the lives of patients, "no matter their creed, nationality, or race," was the supreme concern of physicians and an important step toward creating a thriving nation.[41] Other doctors did not attempt to conceal their efforts to apply social considerations to their medical pursuits. For Dr. José Terrés, uncovering the social status of patients was imperative because it influenced how he treated their illnesses. For example, in the case of tuberculosis, Terrés argued that it would be far more difficult to cure someone from the lower classes because it was nearly impossible for them to meet the minimum dietary requirements needed for a full recovery.[42] This theory influenced how Terrés taught his students to conduct physical exams. As part of the formal examination of patients in hospitals, medical students were supposed to ask each patient a list of questions.[43] These included asking for basic information such as the patient's name, age, marital status, history of illness and hereditary diseases, occupational status, living arrangements, customs, and habits, especially their history with alcohol.[44] In some cases, the background information physicians received altered how they treated a patient's illness.

In the syllabus for the National School of Medicine's 1894 external medicine course, the professor warned his students that relieving a patient's pain was the primary concern, especially in the case of injured children. Excessive pain, coupled with shock from the trauma, can kill a patient within a matter of hours.[45] Severely burned patients were extremely hard to calm—flame had eaten away the subcutaneous layers of their skin, leaving behind a jumble of blackened flesh enmeshed with pieces of clothing and pink, unprotected sections of skin. Many burn victims arrived at the hospital haunted by the stench of their own singed flesh and hair.

Panic and shock manifest themselves in a variety of ways: gut-wrenching screams or the inability to speak, heightened alertness or extreme drowsiness, accelerated pulse or slow heart rate. The effects of shock made it difficult for physicians to administer treatments to mend the wounds.[46] Some doctors developed creative methods to calm trauma patients. In 1897, Guillermo Parra advocated the use of hypnosis to pacify shock victims. Traditionally used to treat hysterics, the therapeutic science of hypnosis could help people ignore their pain, Parra claimed, and he further argued that hypnosis could be a viable alternative to anesthesia in surgeries.[47] Working with the same principles as hypnosis, some physicians tried "harmonic therapy," wherein they attempted to prevent shock by bringing orchestral musicians into the hospital to calm patients.[48]

The lingering trauma of fire injury had a lasting effect on how burn victims functioned in society and how they interpreted the safety of their surroundings. Many people who had endured deep and extensive burns developed mental illness that made them fearful of living, going outside, and having doctors touch them. In July 1904, twenty-four-year-old Melesia Flores was admitted to Hospital San Andrés, complaining of heart pain, difficulty breathing, hardening of the abdomen, fever, and diarrhea. Gabriel González Olvera, the student physician responsible for treating Flores, described the patient's suffering and pain in exacting detail. Two months prior to her admission to the hospital, Flores had suffered extensive burns from an oil fire in her kitchen. White-and-pink striated scars created a craggy rind of skin stretching from her right forearm and fist across her elbow, thus preventing her from fully extending her arm. The left arm had fared no better, again with scarring on the forearm and fist. The flames had eaten away at her index and middle fingers, leaving those two fingers immobile and permanently extended. The scarring also crept up both arms and disfigured her chest and neck with irregularly shaped scars. After his initial examination, the physician suspected that Flores's current ailments all stemmed from the untreated burns she endured—specifically, two months of prolonged buildup of pus beneath her burnt skin, which, left untreated, had caused an infection to spread to her lungs, kidneys, liver, and digestive system. The physician delivered the grave prognosis that Flores would die within a matter of days.

Flores's experience of being burned had a profound impact on her mental state. While in the hospital she suffered extreme bursts of delirium and dementia. One doctor said that her delirium manifested itself as a "true phobia, the terror of being burned," leading her to scream wildly about being persecuted by flames.[49] During a routine examination, one doctor reached his hand out to assess her heart rate, at which point Flores imagined that the doctor held a lit torch that she thought he would use to melt more of her flesh. She howled and cried, begging him not to burn her. In his medical report, González Olvera expressed deep sympathy for the tormented young woman, asserting that such extensive burns would certainly leave a deep impression on anyone's spirit, making it difficult for the victim to return to her previous psychological state. For Flores, months after her burns had healed, the act of being admitted to the hospital and having physicians examine her brought all of her fear of fire and memories of being burned to the surface and her psychological distress manifested itself as extreme suspicion of anyone who approached her. A similar case of a burned patient imagining nightmarish scenarios occurred in 1874 when a young boy, delirious from the pain of sizeable second- and third-degree burns, repeatedly told physicians that his entire body was covered in feathers.[50] For trauma victims, the past and the present can become blurred, making it difficult for these patients to leave tragic events in the past, as a memory. Instead, they remain a present reality, something that they confront on a daily basis.

Melesia Flores's condition worsened day by day, leaving her visibly gaunt and with more frequent delirious outbursts. Following her death, students in the medical school's practicum conducted an autopsy and concluded that the cause of death was most likely hepatitis and tuberculosis.[51] Flores died scared and alone in the hospital bed, but medical students, hospital attendants, and physicians remembered her tragic end and retold her story in medical theses and journal articles. Unlike many illnesses that originate internally, burns forever became a part of a victim's body and psyche, as a constant reminder of the fire accident. For some burned individuals, their quality of life severely decreased and they felt ashamed to leave their homes. In 1898, Dr. Martínez Garza reflected on the many burn cases he had witnessed at the hospital. He claimed that people did not understand the daily agony that burn victims endured, how they would never be as happy as they were before the accidents, and how every day they were forced to reckon with the fact that their beauty had been stripped away.[52] The preponderance of unsightly burned skin created a context in which the population actively sought out creams, tinctures, and remedies to improve skin's appearance.

HOME CARE AND HEALING COMMODIFICATION

The construction of hospitals and clinics and the display of healing products in drugstore windows or in newspaper advertisements stamped the city streets with signs of a robust medical establishment. Yet the abundance of both professional physicians and members of the business sector who sold health-related products and services presented contradictory messages about when and where to seek medical attention.[53] Editors of Mexico's premier medical journal, *La Gaceta Médica*, acknowledged the divide between trained physicians and lay healers and renounced the latter as "scourges of humanity, a testimony to the ignorance and concerns of the common people."[54] Such disdain for "illicit" healers did little to discourage residents from seeking this type of healing in plazas and marketplaces around Mexico City. Knowing that there was competition for healing, trained physicians and drug retailers had to immerse themselves in the medical marketplace and use the expanding medium of advertising to attract potential clients, sometimes by discrediting the competition.

In the late nineteenth century, hospitals were seen more as places where people went to die than as places to recover one's good health. This meant that going to the hospital was an enormously frightening experience. In Federico Gamboa's naturalist novel, Santa concedes her inevitable fate and remarks that her life's journey will leave her in the hospital and then the cemetery, suggesting that hospital visits more often than not spelled death sentences for their patients.[55] Even physicians thought of hospitals as vile places full of miasmas. In his thesis on external medicine, Pedro Martínez Garza explained in detail the process for treating certain wounds (lacerations, burns, and gunshots, to name a few) but concluded by saying that even if a doctor did everything correctly the deplorable conditions in hospitals often resulted in infection.[56] M. A. Posalagua, in his 1874 study of the formation of hospitals in the capital, claimed that hospitals collected all the miasmas and mortalities of the city and therefore were incredibly unsanitary, meaning the sick would only get sicker as a result of spending time in them.[57] Furthermore, it was difficult to isolate the contagious patients from the rest when as many as forty beds sat in the same room. A patient might have started to recover from an ailment only to have contracted a far more serious illness such as typhus or tuberculosis while in the hospital.[58] Many of the hygiene initiatives implemented in homes and businesses did not necessarily reach the hospitals. One physician, appalled by the lack of isolation in hospitals, suggested that the government should invest its funds in creating a brigade of physicians to conduct house calls. This would leave sick patients in a comfortable atmosphere, surrounded by friends and family while not infecting anyone else.[59]

Knowing that hospitals did not necessarily bring health, many sick people preferred to treat illnesses on their own and refused to visit physicians or go to hospitals. This often occurred with smaller burns that did not seem immediately life threatening. Ideally, physicians would receive a burned patient almost immediately after the accident occurred, which would allow them to wash the wound carefully, pick out any debris such as charred clothing or dirt, and apply antiseptic to prevent infections. In one illustrative example, one night a twenty-five-year-old baker visited the latrine and as soon as he put his lit torch on the ground beside him an explosion from the release of hydrogen sulfide and ammonia (used for cleaning latrines) hurled him several yards away. The fire left him with second-degree burns extending from his left shoulder blade down the entire length of his back, across his buttocks, and ending on the outer portions of his right thigh. It was not until five days after the explosion that the family of the burned man requested that Dr. Juan Manuel González Urueña come to treat the young man's injuries.[60] Another case of a delayed visit to the doctor occurred when a fifteen-year-old boy waited eight days to get help at the Hospital Juárez for the third- and fourth-degree burns that covered his arms, thighs, and knees. Due to infections that occurred while his wounds were treated at home, the boy eventually contracted tuberculosis and died.[61] In 1908, electrician Pablo D. Castro suffered grave burns when a container of sulfur exploded, burning the left side of his body badly. For more than a week Castro treated his wounds at home, but he finally went to a hospital.[62] A lack of faith in medical care and financial constraints made the prospect of going to the hospital unviable for most burn victims.

Even though licensed physicians were held in high esteem in society, there was a point at which injured people felt more comfortable administering their own treatments than seeking professional advice. Certainly, a fear of hospitals and financial restraints played a role in how sick and injured people chose to care for themselves, yet there was another force that pulled potential patients away from physicians' offices; namely, the presence of thriving medical capitalism in Mexico City, which emboldened the population to treat ailments at home.[63] Retail drug sellers and therapeutic device developers commodified healing and used advertisements, demonstrations, and articles to remind individuals that they had the freedom to make their own healthcare decisions. Physicians attempted to counter this trend by advertising their practices in much the same way that a drugstore would advertise its products. Physicians offered their services in newspaper classified ads. Some enticed clients with free consultations on certain days of the week, while others showcased their credentials and listed where they received their medical degrees.[64] Other physicians mimicked the lofty arrogance found in pharmaceutical advertisements of the time, guaranteeing such things as a total cure for diabetes, obesity, syphilis, or constipation in ten days or less and without invasive surgeries.[65]

Many of these advertisements targeted mothers and wives because of their traditional roles as caregivers. In one article aimed at housewives, the unnamed author claimed that all mothers should have rudimentary medical knowledge to treat their children in the event of an accident and reminded mothers, "In homes with children, there are always burns." The article explained that for burn wounds, a mother should wash and disinfect the burn with a little limewater and baking soda, cover it with a damp cotton cloth, and repeat every twenty-four hours. Only if the pain subsisted for several days was a visit to the hospital or doctor's office deemed necessary.[66] In the "Helpful Recipes" section of *El País*, women were encouraged to make paste solutions out of salt, water, and baking soda to relieve the pain of burns, whereas a similar column in *El Diario* recommended putting ground coffee on third-degree burns.[67] All of these suggested simple treatment options made of ingredients that could be found in a typical kitchen that anyone could administer with ease.

Published records suggest that home treatments often ended badly and were often accompanied by commentary from a licensed physician urging people not to self-diagnose or self-treat their ailments. In one such case, police reports indicate that after María Guadalupe Delgado's dress caught fire, leaving her with extensive burns on her hands, legs, face, and chest, she tried to cure her wounds by herself. Despite thoroughly cleansing and disinfecting the burns at home, her home treatments were unsuccessful and she died suffering in her house.[68]

In 1906, a mother took her young daughter to a *curandera* (healer) to receive a treatment for an upset stomach, but the substance that the young girl ingested poisoned her and took her life. This case led physicians to warn patients of seeking advice from healers who did not have official medical training.[69] Based on the recorded interactions between physicians and patients it is clear that seeking professional medical help was usually a last resort, and more often than not burn victims would treat themselves or employ the abundance of over-the-counter options to cure their ailments.

Just as Mexico's political elite valued European political ideologies and academic training, health consumers also placed value on ointments and elixirs imported from Europe. In their attempts to attract a Mexican clientele, the Holloway Ointment Company advertised, "Nearly every hospital in Europe uses this remedy to cure all external ailments," and while the advertisement never explained the ointment's ingredients, it claimed to have magical healing properties.[70] Other companies also heralded the effectiveness of their ointments and elixirs by emphasizing their European roots, such as the "Great German Remedy" that cured everything from sciatica to gout and promised to relieve any pain immediately.[71] Grandiose claims found in advertisements for "Dr. Robinson's Ointment" promised that after applying his ointment customers could watch ailments heal before their eyes. These advertisements reassured the population that home treatment was a viable option.[72]

Some of these treatments ignored the newest European solutions and instead looked to traditionally indigenous healing methods. One company harnessed the healing energy of "tree sap that the Indians call Aporó," which advertisements claimed could immediately cure wounds, burns, contusions, and other external ailments. For burns, the advertisement promised that Aporó would not leave any scarring.[73] Tepezcohuite, the bark of the *Mimosa tenuiflora* tree, was also an option for treating burns. The *Mimosa tenuiflora* grows in limited regions of highland Chiapas, and Maya communities have used its bark for healing purposes for several centuries. Toasting the tepezcohuite bark and grinding it into a powder unlocks the bark's medicinal properties, and in this form it could easily be mixed with liquids and applied directly to burnt skin. At the turn of the twentieth century several Mexican scientists were aware of the existence of tepezcohuite, yet its use as a treatment for burns never came up in official medical documents.[74]

One indigenous plant that bridged the divide between folk remedy and modern pharmaceutical was matarique (or *maturí*), a plant native to Sonora and Chihuahua. In 1888, Fernando Güereña applied to patent the process of extracting the juices from the root of the matarique for use in a topical ointment that had a strong healing power on injured flesh. Despite explaining that the Yaqui Indians had long employed the matarique root to heal wounds and relieve pain, Güereña decided it was most appropriate to call it the "Güereña Root" after himself.[75] Aside from one or two vague references to the Yaqui in official documents, including one mention that "the Yaquis call the plant *Maturí*, a word that reputedly means pain killer," Güereña and scientists erased the vital role that the Yaquis played in bringing the healing properties of the matarique to a larger public.[76] Instead, all credit went to Güereña and the scientists who introduced matarique to the international medical community and explained its healing properties in scientific terms.

Güereña's patent request drew considerable attention and within a year the Ministry of Development commissioned the National Medical Institute to conduct a thorough investigation of the plant's chemical properties and its effect on the human body. Dr. Manuel Urbina, the official botanist of the National Medical Institute, who was responsible for collecting and classifying plants, studied the matarique plant and published a detailed report of his findings. Using modern plant taxonomy, Urbina measured the leaves and roots, examined the reproductive organs, and noted the characteristics of the seeds and flowers to determine the matarique's genus and species, which he ultimately classified as *Cacalia cerviariaefolia*.[77] Employing a globally acceptable scientific language, and without taking into consideration Yaqui experiences and knowledge systems tied to matarique, Urbina transmitted the existence of a specifically Mexican plant to a wider medical botany community. The chemist commissioned to analyze the chemical properties of the plant also contributed to the globalization of medical knowledge. Donaciano Morales brought samples of the root with him to England, where he tested whether or not it contained alkaloids and if it was alcohol soluble. Working in English laboratories alongside English chemists, Morales created a scientific dialogue that challenges assumptions that European scientific knowledge was imposed on Mexico, yet it still shows that it took an English colleague and laboratory to get anyone to pay attention, and that the Mexican-European scientific nexus existed, albeit in an unequal way. After Urbina and

Morales published their findings about matarique, other physicians and scientists cited the investigation widely in international journals.[78]

Although Morales hypothesized that a root with those particular chemical properties could have antipyretic (antifever) and antidysentery qualities, neither of the aforementioned experiments addressed the effects of matarique on the body. A team of faculty and students from Mexico's National Medical School, led by Dr. Francisco Rio de la Loza, conducted a set of physiological experiments to understand matarique's potential for application. First, they injected a frog with fifty centigrams of the matarique extract, which resulted in paralysis, strained respiratory function, and finally death after eight hours. After altering the dosage slightly, the doctors injected two more frogs with the extract and saw improved results. They then tried the same experiment on a dog and observed an irregular heartbeat, almost undetectable blood pressure, muscle relaxation, gurgling noises, bloating, constipation, dilated pupils, and extreme fatigue. The next day the dog had completely recovered. In subsequent therapeutic applications on human subjects, doctors noted that the topical application of matarique reduced pain and helped heal wounds quickly. Suspecting that it acted like an antiseptic, the doctors praised Güereña for discovering the remedy and encouraged doctors to prescribe two teaspoons of the extract per day to patients suffering ailments ranging from nervous disorders to hemorrhoids to hemophilia.[79]

Güereña wanted to profit from matarique's healing properties, but in order to get official recognition he had to have experts, including botanists, chemists, and physicians, validate the claims he made in his patent request. The matarique case explicates the complex process of getting medicine approved through official, state-sanctioned channels in order to market the product domestically and internationally. While Mexicans had used matarique for centuries to treat burns and other skin ailments, the matarique root acquired a luster of authenticity once authorities tested it and agreed to patent it. Here the medical appropriation of matarique fits into a broader pattern of modernization. It is at once a story of silencing Yaqui voices, appropriating their knowledge, and then tossing that material into the Mexican-European science community.[80]

Home and professional treatments for burns offer only a sample of how officials and citizens folded fire and fire-related illnesses into the broader spectrum of health in the city. With the advent of the Bacteriological Revolution in the 1880s, hygienists applied sanitation codes in an attempt to decontaminate the city and cleanse it of risk. Bacteriology, and particularly Louis Pasteur's germ theory, equally altered how physicians treated the accidents of industrial development. Understanding that burned skin easily succumbed to infection, physicians applied disinfectants and sterilized surgical tools, which significantly decreased the likelihood of infection. Burn treatment specializations at the medical school and experiments with reconstructive surgery inspired pharmacists, inventors, and housewives to experiment on their own. Marketplaces and plazas full of readily available ointments and elixirs to cure burns and repair flesh ultimately forced burn victims to make difficult decisions about who to entrust with life-threatening ailments and where to receive medical treatment. In the case of the matarique root, the line between modern and traditional became blurred when trained scientists in Mexico and Europe appropriated the knowledge of the root's health benefits. Regardless of the important role that local knowledge played in treating burns with matarique, the local healers who applied their experience and knowledge for patients' benefit were not seen as contributors to the global medical community.

CONCLUSION

Fire's survival requires that it gnaw away at the materials that make up the city—wood, cloth, floorboards, and shingles. Its unrelenting search for fuel can make its behavior erratic and cause it to spread quickly and without respect for traditional property boundaries or social divisions. Due to fire's indiscriminate destructiveness, it cannot easily be categorized as an individual danger. Or, as one underwriter aptly put it, fire is "more than a private misfortune. It is a public dereliction."[1] Controlling a hazard like fire that exists in both private and public spaces requires the participation of individuals and the community as a whole. This fact made fire control at the turn of the twentieth century in Mexico City ill fitted to the traditional liberal mold that emphasized individualism, self-improvement, and private property. Through decades of experimentation with ways to create a safer city, citizens discovered that controlling, managing, and preventing fire required action at all levels, ranging from sweeping, citywide programs initiated by public servants to narrow, home-centered consumer practices of individuals.

Fire hazards illuminate a fundamental tension in this period of Mexican history; namely, that the contradictory processes of increasing privatization of responsibility and increasing public intervention both occurred at this time. By examining this tension through the lens of science and technology, it becomes clear that neither the liberal spirit of individualism nor excessive reliance on the public sector fully predominated at this moment. Instead, these seemingly divergent ways of approaching urban problems reinforced one another. By examining Mexico City's fire problem, we see that both individualistic and collective approaches to fighting fire ultimately relied on the application of science to subdue flames, and it is through the science of regulation and the application of technology that the competing worlds of individual and governmental responsibility connect.

The way Mexico City residents responded to increased instances of fire represents a vision of modernity and human progress that was rooted in technological innovation, scientific expertise, and individual ingenuity to overcome obstacles. As a result, this intellectual environment gave rise to new fire-related occupations and specialized knowledge in both the private and the public spheres, all of which fundamentally shaped how residents interacted with their city. New public servants, among them firemen, city inspectors, and municipal engineers, emerged alongside private entrepreneurs, including inventors, medical vendors, and insurance agents. Collectively, they represented a positivist faith in the idea that the knowledge acquired by experts could be applied to solve problems related to fire. When individuals and communities equally accepted responsibility for preventing and controlling fire, the city and its citizens were at their safest because dangers were being mitigated at all levels. When this balance of responsibility shifted in either direction, the damage that a fire could cause increased.

Occasionally, the rise of fire expertise inadvertently increased risk. In his book on the development of disaster expertise, historian Scott Gabriel Knowles explains, "In their work towards ending disasters, the experts have supplied much of the confidence necessary to take bigger risks," which had the unintended effect of increasing danger.[2] Even though Knowles was primarily referring to other historical processes, including the quest to construct taller skyscrapers because of improved construction technologies and the decision to build homes in floodplains because insurance companies have agreed to cover damage, his general argument is effective here as well. An illustrative example of Mexico City's transition to expert authority involves the professionalization of the fire brigade. In mid-nineteenth-century Mexico City a fire scene would have included neighbors lining up to form a bucket brigade to extinguish the flames by themselves. By the early twentieth century, the scene of a fire most likely included bystanders waiting around for the professional fire brigade to arrive. By waiting for professionals to arrive, the onlookers validated the authority of the brigade and put an incredible degree of faith in the firemen's tools, equipment, and training. Insurance agents complained incessantly about how this faith in the fire department led some citizens to feel that their efforts to form a bucket brigade would be useless.[3] Trusting that experts and technology would save the day meant that small fires that could have been extinguished easily with buckets of water became much larger and uncontrollable as spectators waited for professional support. Conversely, there were several instances when bystanders wanted to help at the scene of a fire, but firemen asserted their authority by physically holding back the onlookers. At one of the Volador market fires, firemen used bayonets to threaten observers who got too close to the flames, thus signaling that trained experts had the fire under control.[4]

This reliance on science and technology helped create a general attitude that entrusted experts to innovate their ways out of problem situations. Some citizens had the conviction that experts would use their knowledge to fix problems that plagued the city, which in some cases led to almost an aversion to fire safety methods that did not involve some form of technology. Engineer-inspectors saw this phenomenon firsthand and were astounded that so many business owners refused to follow city ordinances and fire codes because they felt the new hydrants or sprinklers provided enough protection. After a 1905 fire at the Teatro Colón, city engineer Abraham Chávez wrote a heated letter to the Ayuntamiento outlining the many deficiencies in public hygiene and safety in the capital. He noted that theater owners frequently disregarded safety regulations such as the ban on smoking cigarettes or the prohibition of certain stage props, and that during opening night of the Opera Mignon, a candle fell over on the stage and ignited the curtain. Despite clear regulations that prohibited candles on theater stages, the theater owner and the stage actors decided not to comply.[5] Theater owners were especially negligent when it came to following city fire codes that restricted occupancy or called for the installation of nonflammable building materials, which in both cases cut into the owner's profits. Ultimately engineers dealt with neglectful theater owners by crafting specific theater regulations and initiating mandatory monthly inspections.

Public health officials faced similar resistance as they tried to make health a reality for more residents by applying preventative approaches that regulated behaviors. Actions such as smoking cigarettes in factories or improperly disposing of oily rags now violated the city's fire and sanitation codes. These various approaches to regulating the city attempted to improve the human condition and make a healthy and productive population. Regardless of fire codes and regulations, residents continued to equate protection with technology, even as inspectors and insurers continued to express that "an ounce of prevention is better than a pound of cure."[6] In part, the "cure" of purchasing fire control devices replaced preventative techniques because shooting a fire extinguisher, flinging a hand grenade, or hearing the roaring sirens from a fire engine evoked excitement that preventative methods could never match.

Burgeoning consumer culture and advertising convinced capital residents that purchased protection was the best form of protection, and evidence of this sentiment infiltrated all levels of society. Ayuntamiento officials measured their success at dealing with fire hazards by proudly listing off the number of hoses, fire engines, ladders, helmets, and extinguishers available to safeguard the city. Conversely, these officials also measured their failures at dealing with fire hazards by listing off the available equipment or comparing Mexico City's lack of equipment to that of other modern cities. On the level of the individual, purchased goods helped define one's social status, and in the case of fire protection purchases, helped define one's status as a concerned and prudent citizen. The visibility of hanging a fire grenade on a living room wall or attaching a water cistern for the sprinkler system to the roof of a building made an individual's devotion to safety discernable in a way that seemingly antiquated preventative techniques could not. Simple precautions like not storing an excessive amount of flammable substances in central areas or remembering to close windows when there were lit candles inside to avoid accidentally spreading a flame with wind did not have the same appeal that purchased protection did.

For preventative, nontechnological methods to work, entire neighborhoods and communities had to abide by them. For example, it did little good for one household to get rid of its supplies of flammable substances if all the surrounding households kept theirs. Instead of relying on neighbors to cooperate with one another by adopting common fire prevention techniques, city residents increasingly relied on technology they could purchase. Fire extinguishers, fire alarms, or automatic sprinklers gave residents peace of mind that, if a fire were to erupt, they could manage to extinguish the flames on their own. This insistence on self-reliance reflected the liberal ethos of the time. Thus, purchasing a safety device or an insurance policy provided a personal sense of security and also was ideologically consistent with liberal precepts. However innovative these technologies may have been, devices and chemicals could not always stop the spread of fire, nor could ever more complex insurance policies compensate for the material or human losses caused by fire.

In an industrial fire regime characterized by more explosions and hotter burning temperatures due to the presence of combustibles, fires plagued the city and the haunting aftermath of flame and smoke left both buildings and bodies scarred. Blackened buildings could be torn down and replaced, but the loss of life was permanent. Just as other areas of fire-related specialization emerged, public servants and private entrepreneurs adapted their professions to confront the industrial fire regime. They applied the latest approaches in germ theory, sanitation, and disinfection to fire-related ailments that affected both the body and the city. Yet community members struggled to determine which groups of health experts to trust. Mothers, wives, and marketplace vendors had traditionally cared for everyday burns, but as oils and gases became more common, fire ate away at skin more quickly and the everyday burns became deeper and harder to heal. The alternative to home remedies involved visiting a physician's office or hospital, yet that option was costly and unfeasible for many residents.

The ability of some people to purchase protection while others could not was emblematic of the deep inequalities that characterized this era in Mexican history. Unequal distribution of power, resources, and in this case safety became a hallmark of the late nineteenth and early twentieth centuries, and opposition to such stark inequities eventually became one of the rallying cries of the 1910 revolution. But well before the onset of revolutionary upheaval, many of the city's residents sought to create a more equitable and responsible way of distributing safety. Public servants, particularly engineer-inspectors hired by the municipal government, were some of the first people in the city to acknowledge openly that fire protection should be extended more equitably, since fire threatened all social classes. Abraham Chávez was one of the more vocal advocates for redesigning fire safety protocols to benefit all residents. As city engineer, Chávez noted that engineer-inspectors were "genuine representatives of all citizens," and he argued that it was their duty to protect everyone from the evils of fire and "to monitor closely their woes and attempt to fight with all our strength to improve them." According to Chávez, it was in fire hazards that he "found a real social wound that is eating away at our society, undermining its morality," and only through equitable distribution of support and protection, in the form of both government regulation and personal investment, could the city expect to see improvement.[7]

It is not surprising that the engineers' demands for more equitable distribution of resources and protection did not resonate with most government officials. On an ideological level, the engineers' appeals marked a major deviation from the Porfirian mindset, which valued ideas of elite progress often at the expense of the needs of the poor. But Chávez was a trained expert who studied how fire functioned, seeing that it had no allegiances to social boundaries and that if it erupted in a lower-class neighborhood there was no assurance that it would not drift over and across neighborhood borders. In effect, by observing fire's behavior he knew that fire could affect anyone at any time, and for this reason he advocated for systematic change in the approach to this menacing urban problem.

By 1912, in the midst of revolutionary struggle, Chávez had aligned with the social goals of the revolutionaries and used his official post to advocate for more social equality. Even though Chávez never outwardly opposed the Porfirian government or officially backed the revolutionaries, in his proposition to expand fire inspection he made an impassioned case for protecting all citizens, explaining that public servants needed to refocus their attention and assist "the proletariat and even the middle class, so that they are equally guaranteed their lives and interests." Through protection, he argued, poor citizens would be given an equal opportunity to fulfill both their personal goals and their civic duties without worrying about the "true gore" of fire.[8] By the time of the revolution, the definition of citizenship came to encompass the fundamental rights to property ownership, productivity, health, and safety, all of which could be affected by the presence of unruly fire.

NOTES

INTRODUCTION. MODERNITY AND ITS ACCIDENTS

1. Environmental historian Stephen Pyne suggests that fire has had such a profound impact on human development that "the Anthropocene might equally be called the Pyrocene." See "The Fire Age," *AEON Magazine*, May 5, 2015, https://aeon.co/essays/how-humans-made-fire-and-fire-made-us -human.

2. Ulrich Beck refers to this as "reflexive modernity" in his book, *Risk Society: Towards a New Modernity* (London: Sage, 1992), 17–50.

3. Mauricio Tenorio-Trillo, *I Speak of the City: Mexico City at the Turn of the Twentieth Century* (Chicago: University of Chicago Press, 2012), xxiii.

4. Stephen J. Pyne, *World Fire: The Culture of Fire on Earth* (Seattle: University of Washington Press, 1997), 23–25. Scott Gabriel Knowles refers to this period as the "Conflagration Era"; see *The Disaster Experts: Mastering Risk in Modern America* (Philadelphia: University of Pennsylvania Press, 2011), 15.

5. Germán Vergara, "Energy, Environment, and Society in the Basin of Mexico until the Nineteenth Century," in *Mexico in Focus: Political, Environmental and Social Issues*, ed. José Galindo (New York: Nova Science, 2015), 4–5.

6. Matthew Vitz, "'To Save the Forests': Power, Narrative, and Environment in Mexico City's Cooking Fuel Transition," *Mexican Studies/Estudios Mexicanos* 31, no. 1 (2015): 127.

7. Vera S. Candiani, *Dreaming of Dry Land: Environmental Transformation in Colonial Mexico City* (Stanford, CA: Stanford University Press, 2014); Germán Palacio Castañeda, *Historia ambiental de Bogotá y la Sabana, 1850–2005* (Amazonas, Colombia: Universidad Nacional de Colombia, Sede Amazonia, 2008); Lise Sedrez, "Urban Nature in Latin America: Diverse Cities and Shared Narratives," in *New Environmental Histories of Latin America and the Caribbean*, ed. Claudia Leal, José Augusto Pádua, and John Soluri, Rachel Carson Center Perspectives 7 (Munich: Rachel Carson Center, 2013), 59–66; Lise Sedrez, "Latin

American Environmental History: A Shifting Old/New Field," in *The Environment and World History*, ed. Edmund Burke III and Kenneth Pomeranz (Berkeley: University of California Press, 2009), 255–75; Richard C. Hoffman, Nancy Langston, James McCann, Peter Perdue, and Lise Sedrez, "AHR Conversation: Environmental Historians and Environmental Crisis," *American Historical Review* 113, no. 5 (2008): 1431–65.

8. Stuart McCook, "Focus: Global Currents in National Histories of Science: The 'Global Turn' and the History of Science in Latin America," *Isis* 104, no. 4 (2013), 776; Gabriela Soto Laveaga, *Jungle Laboratories: Mexican Peasants, National Projects, and the Making of the Pill* (Durham, NC: Duke University Press, 2009), 6–8; Katherine Elaine Bliss, *Compromised Positions: Prostitution, Public Health, and Gender Politics in Revolutionary Mexico City* (University Park, PA: Pennsylvania State University Press, 2001).

9. J. R. McNeill, *Mosquito Empires: Ecology and War in the Greater Caribbean, 1620–1914* (New York: Cambridge University Press, 2010), 2; Mariola Espinosa, *Epidemic Invasions: Yellow Fever and the Limits of Cuban Independence, 1878–1930* (Chicago: University of Chicago Press, 2009); Elizabeth A. Fenn, *Pox Americana: The Great Smallpox Epidemic of 1775–82* (New York City: Hill and Wang, 2001).

10. This approach draws on Timothy Mitchell's proposal that scholars go beyond questions of human motive or human interactions to understand social change. In particular he considers how the interactions between mosquitos, parasites, chemicals, and water sources contributed to Egypt's twentieth-century experience. See "Can the Mosquito Speak?" in *Rule of Experts: Egypt, Techno-Politics, Modernity* (Berkeley: University of California Press, 2002), 30–31 and 46.

11. In her article, Regina Horta Duarte explains that access to valuable resources like potable water often depended on class-based spatial configurations. See "'It Does Not Even Seem Like We Are in Brazil': Country Clubs and Gated Communities in Belo Horizonte, Brazil, 1951–1964," *Journal of Latin American Studies* 44, no. 3 (2012): 446.

12. Tenorio-Trillo, *I Speak of the City*, 307.

13. Mónica Blanco and María Eugenia Romero Sotelo, "Cambio tecnológico e industrialización: La manufactura mexicana durante el Porfiriato (1877–1911)," in *La industria mexicana y su historia: Siglos XVIII, XIX, y XX*, ed. María Eugenia Romero Sotelo (Mexico City: DGAPA-FE-UNAM, 1997); Dawn Keremitsis, *La industria textil mexicana en el siglo XIX* (Mexico City: SepSetentas, 1973); Stephen H. Haber, *Industry and Underdevelopment: The Industrialization of Mexico, 1890–1940* (Stanford, CA: Stanford University Press, 1989); John Coatsworth, *Growth against Development: The Economic Impact of Railroads in Porfirian*

Mexico (Dekalb: Northern Illinois University Press, 1981); Marvin D. Bernstein, *The Mexican Mining Industry, 1890–1950: A Study of the Interaction of Politics, Economics, and Technology* (Albany: State University of New York Press, 1964).

14. See Edward Beatty, *Technology and the Search for Progress in Modern Mexico* (Berkeley: University of California Press, 2015); Steven B. Bunker, *Creating Mexican Consumer Culture in the Age of Porfirio Díaz* (Albuquerque: University of New Mexico Press, 2012), for information on advertisements and the promises of technology; Rubén Gallo, *Mexican Modernity: The Avant-Garde and the Technological Revolution* (Cambridge, MA: MIT Press, 2010); Michael Matthews, *The Civilizing Machine: A Cultural History of Mexican Railroads, 1876–1910* (Lincoln: University of Nebraska Press, 2013); Araceli Tinajero and J. Brian Freeman, eds., *Technology and Culture in Twentieth-Century Mexico* (Tuscaloosa: University of Alabama Press, 2013).

15. Erica Berra Stoppa, "La expansión de la Ciudad de México y los conflictos urbanos, 1900–1930" (PhD thesis, Colegio de México, 1983), 88; Jesús Galindo y Villa, *Reseña histórica-descriptiva de la Ciudad de México que escribe el Regidor del Ayuntamiento, por encargo del Señor Presidente de la misma corporación D. Guillermo Landa y Escandón, y expresamente para los delegados a la segunda conferencia internacional americana* (Mexico City: Díaz de León, 1901), 7–8.

16. François-Xavier Guerra, *México: del antiguo régimen a la revolución* (Mexico City: Fondo de Cultura Económica, 1988), 1:338; Secretaria de Económica, *Estadísticas sociales del porfiriato, 1877–1910* (Mexico City: Dirección General de Estadística, 1956), 73.

17. Alejandra Moreno Toscano y Enrique Florescano, *El sector externo y la organización especial y regional de México (1521–1910)* (Mexico City: UAP, 1970), 49; Pablo Macedo, *La evolución mercantil: comunicaciones y obras públicas, la hacienda pública* (Mexico City: J. Ballescá, 1905).

18. Jonathan Kandell, *La Capital: The Biography of Mexico City* (New York: Random House, 1988), 331–33.

19. Gustavo Garza Villarreal, *El proceso de industrialización en la Ciudad de México (1821–1970)* (Mexico City: El Colegio de México, 1985), 103–4; Sergio de la Peña, *La formación del capitalismo en México* (Mexico City: Siglo XXI Editores, 1975), 51.

20. Christian Brannstrom's revisionist account of the so-called wood hypothesis proposed by Warren Dean found that, similar to central Mexico at the turn of the twentieth century, São Paulo also relied on a combination of energy sources that included biomass, hydroelectricity, and fossil fuels. See "Was Brazilian Industrialisation Fuelled by Wood? Evaluating the Wood Hypothesis, 1900–1960," *Environment and History* 11, no. 4 (2005): 395–430.

21. Kandell, *La Capital*, 346.

22. José Luis Blasio, *Maximilian, Emperor of Mexico: Memoirs of His Private Secretary*, trans. Robert Hammond Murray (New Haven, CT: Yale University Press, 1934), 53.

23. Alain Corbin, *El perfume o el miasma: el olfato y lo imaginario social, siglos XVIII y XIX*, Sección de obras de historia (Mexico City: Fondo de Cultura Económica, 1987), 105–21; Marcos Arróniz, *Manual del viajero en México* (1858; repr., Mexico City: Instituto Mora, 1991), 117.

24. Silvia Marina Arrom, *Containing the Poor: The Mexico City Poor House, 1775–1871* (Durham, NC: Duke University Press, 2000), 241.

25. Información gubernativa sobre las causas que ocasionaron el incendio de la Plaza del Volador, April 23, 1871, Archivo Histórico del Distrito Federal [hereafter AHDF]: Ayuntamiento Gobierno del Distrito Federal [hereafter AGDF], Rastros y Mercados, vol. 3733, exp. 518, fs. 45–46.

26. Incendio ocurrido en la panadería del Pte. Quebrado No. 7, April 20–25, 1907, AHDF: AGDF, Gobierno del Distrito, Fábricas, vol. 1604, exp. 356, fs. 1–4.

27. Albino G. Serrano, Comandante del Cuerpo de Bomberos, March 4, 1908, Archivo General Municipal de Puebla [hereafter AGMP], tomo 408, ficha 16526, legajo 23, fs. 49–82. During this period, matches that were not coated with varnish easily caught fire when in contact with dry air or high temperatures; see Archivo General de la Nación [hereafter AGN], Francisco Tadeo Linder, "Fabricación de cerillos, fósforos y yesca de seguridad," Patentes y Marcas, caja 5, exp. 369, November 27, 1858; AGN, Agustín Rousscav y Luís Chaubet, "Fabricación de fósforos y cerillos inofensivos," Patentes y Marcas, caja 4, exp. 366, November 3, 1858.

28. M. Nafera to S. Alcalde Municipal, March 23, 1866, AHDF: AGDF, Policía, Incendios, vol. 3649, exp. 59, fs. 23.

29. Charles F. Walker, *Shaky Colonialism: The 1746 Earthquake-Tsunami in Lima, Peru, and Its Long Aftermath* (Durham, NC: Duke University Press, 2008); Mark A. Healey, *The Ruins of the New Argentina: Peronism and the Remaking of San Juan after the 1944 Earthquake* (Durham, NC: Duke University Press, 2011); Jürgen Buchenau and Lyman L. Johnson, *Aftershocks: Earthquakes and Popular Politics in Latin America* (Albuquerque: University of New Mexico Press, 2009).

30. Greg Bankoff, *Cultures of Disaster: Society and Natural Hazard in the Philippines* (London: RoutledgeCurzon, 2003), 14–17.

31. American School of Correspondence, *Cyclopedia of Fire Prevention and Insurance: A General Reference Work, Prepared by Architects, Engineers, and Practical Insurance Men*, vol. 1 (Chicago: American School of Correspondence, 1912), 18.

32. Ann S. Blum, "Conspicuous Benevolence: Liberalism, Public Welfare, and Private Charity in Porfirian Mexico City, 1877–1910," *Americas* 58, no. 1 (2001): 8–9.

33. Julia Rodríguez, *Civilizing Argentina: Science, Medicine, and the Modern State* (Chapel Hill: University of North Carolina Press, 2006), 3.

34. In her study of modernity during Argentina's "golden era" from the 1880s to 1914, Julia Rodríguez highlights many of the tensions and incongruities of liberalism that can be seen in both Argentina and Mexico, explaining that "liberalism did not bring progress, let alone freedom and equality, for all." *Civilizing Argentina*, 3. See also Leopoldo Zea, *El positivismo en México: nacimiento, apogeo y decadencia* (Mexico City: Fondo de Cultura Económica, 1968). Alan Knight uses the term "developmentalist liberalism" to describe the Porfirian political elite and *científico* outlook that equated liberalism with material progress. See *The Mexican Revolution* (Cambridge: Cambridge University Press, 1986), 69; as well as Charles A. Hale, *The Transformation of Liberalism in Late Nineteenth-Century Mexico* (Princeton, NJ: Princeton University Press, 1989), 3 and 20–21.

35. Juan de Dios Peza, *La beneficencia en México* (Mexico City: Imprenta de Francisco Díaz de León, 1881), 33–34.

36. Moisés Gonzalez Navarro, *La pobreza en México* (Mexico City: Colegio de México, 1985), 54–56.

37. This introduced an astounding number of city improvement projects, including the expansion of green spaces and the construction of the Palace of Fine Arts (Palacio de Bellas Artes). Federico Fernández-Christlieb, *Mexico, Ville Néoclassique: Les espaces et les idées d'aménagement urbain (1783–1911)* (Paris: L'Harmattan, 2002), 22; Michael Johns, *The City of Mexico in the Age of Díaz* (Austin: University of Texas Press, 1997), 17–21; Emily Wakild, "Naturalizing Modernity: Urban Parks, Public Gardens and Drainage Projects in Porfirian Mexico City," *Mexican Studies/Estudios Mexicanos* 23, no. 1 (2007): 101–23.

38. William H. Beezley, *Judas at the Jockey Club and Other Episodes of Porfirian Mexico*, 2nd ed. (Lincoln: University of Nebraska Press, 2004), 14.

39. Bliss, *Compromised Positions*, 7; Johns, *City of Mexico*, 17–21.

40. Garza Villarreal, *El proceso de industrialización en la Ciudad de México*, 117–22.

41. Joseph Bird, *Protection against Fire, and the Best Means of Putting Out Fires in Cities, Towns, and Villages, with Practical Suggestions for the Security of Life and Property* (New York: Hurd and Houghton, 1873), 112.

42. Consejo Superior de Salubridad to Presidente del Consejo Superior de Gobierno del Distrito Federal México, April 4, 1910, AHDF: AGDF, Policía, vol. 617, exp. 37.

43. Franz Mauelshagen, "Disaster and Political Culture in Germany since 1500," in *Natural Disasters, Cultural Responses: Case Studies toward a Global Environmental History*, ed. Christof Mauch and Christian Pfister (Lanham, MD: Lexington Books, 2009), 58.

44. The population from 1876 to 1910 quintupled; see Ramona Pérez Bertruy, "La constitución de paseos y jardines públicos modernos en la Ciudad de México durante el Porfiriato: Una experiencia social," in *Los espacios públicos de la ciudad siglos XVIII y XIX*, ed. Carlos Aguirre Anaya, Marcela Dávalos, María Amparo Ros (Mexico City: Casa Juan Pablo, 2002), 314–34.

45. Secundino E. Sosa, "Su higiene, sus enfermedades" (MD thesis, La Escuela Nacional de Medicina, 1888), 34.

CHAPTER ONE. FIGHTING FIRE, FIGHTING FEAR

Epigraph: Henry L. Champlin, *The American Firemen: Essays, Lurid Leaves, Sketches, Sparks: A Standard Work on Fire Matters* (Boston: H. L. Champlin, 1875), 7.

1. Incendio en la Calle San Antonio Abad, February 10, 1866, AHDF: AGDF, Policía, Incendios, vol. 3649, exp. 59, fs. 8.

2. Palacio Municipal, February 26, 1866, AGN: Segundo Imperio, vol. 136, caja 38, exp. 21.

3. M. Nafera to S. Alcalde Municipal, March 23, 1866, AHDF: AGDF, Policía, Incendios, vol. 3649, exp. 59, fs. 23.

4. "El Bombero," *El Monitor Republicano*, December 9, 1883, 2. In her analysis of the sounds of the Overland Trail, Sarah Keyes examines how "sounds and silences possessed the power to comfort and distress"; see "'Like a Roaring Lion': The Overland Trail as a Sonic Conquest," *Journal of American History* 96, no. 1 (2009): 23.

5. "Un torbellino de fuego destruyo anoche 'El Palacio de Hierro,'" *El Imparcial*, April 16, 1914, 1.

6. "Un voraz incendio convierte en escombros una fotografía," *El País*, January 7, 1909, 1.

7. *Voz de México*, June 13, 1890, 2.

8. Maurice Halbwachs, *On Collective Memory* (Chicago: University of Chicago Press, 1992), 38–40; Lucien Febvre, "Sensibility and History: How to Reconstitute the Emotional Life of the Past," in *A New Kind of History: From the Writings of Febvre*, ed. Peter Burke, trans. K. Folca (London: Routledge and Kegan Paul, 1973), 14–15.

9. In his study of crime in Porfirian Mexico City, James Garza explains how fear created the imagined underworld. See *The Imagined Underworld: Sex, Crime, and Vice in Porfirian Mexico City* (Lincoln: University of Nebraska Press, 2007), 3; William E. French, "Imagining and the Cultural History of Nineteenth-Century Mexico," *Hispanic American Historical Review* 79, no. 2 (1999): 249–67.

10. "El fuego en nuestros teatros," *El Monitor Republicano*, May 31, 1887, 3.

11. "Terrible catastrofe: San Francisco convertido en ruinas," *El Imparcial*, April 19, 1906; "Las ruinas de San Francisco: relato de un testigo presencial," *El Imparcial*, April 20, 1906; "El Incendio Cesa: comienzan á llegar auxilios," *El Imparcial*, April 21, 1906; "La reconstrucción de San Francisco: límites exactos de la zona destruido," *El Imparcial*, April 22, 1906; "San Francisco renace: ya se levantan nuevos muros," *El Imparcial*, April 24, 1906.

12. Broadside engraving, José Guadalupe Posada, "Espanto Teremotos y Formidable Incendio en San Francisco California! ¡La Ciudad Entre Las Llamas!—Millares de Victimas!—¡Catastrofe Nunca Vista!" (Mexico City: Antonio Vanegas Arroyo, 1906), Jean Charlot Collection, University of Hawaii at Manoa Library.

13. "Incendio de una casa de locos," *El Siglo Diez y Nueve*, June 6, 1890, 2; "Detalles de un incendio," *Voz de México*, June 10, 1890, 1.

14. "El incendio en Chicago, golpe de vista," July 23, 1893; "Incendio de la exposición de Chicago," and "Incendio en Chicago," *Monitor Republicano*, May 14, 1885, 4.

15. D. Antonio Martínez de Castro, "Reclamación de daños y perjuicios ocasionados por el incendio. Derechos recíprocos de los coinquilinos. Prestación y prueba de la culpa. Estudio comparado de diversas legislaciones," *Gaceta de los Tribunales de la Republica Mexicana*, March 8, 1862, vol. 3, no. 10, 186–96.

16. "Incendio en una coheteria," *El País*, October 13, 1903, 2.

17. Jon Bannister and Nick Fyfe explain that one of the reasons people are plagued by fear in cities is due to the prospect of victimization, especially in cases of crime; see "Introduction: Fear and the City," *Urban Studies* 38, no. 5–6 (2001), 809.

18. Aviso al Publico por José María Icaza, June 5, 1842, AHDF: AGDF, Policía, Incendios, vol. 3649, exp. 44.

19. Broadside engraving, José Guadalupe Posada, "Ejemplar y ciertísimo suceso en la República Mexicana. Las verdaderas causas de temblor del día 2 de noviembre de 1894." Fernando Gamboa Collection of Prints by José Guadalupe Posada, Center for Southwest Research, University Libraries, University of New Mexico.

20. Juan Manuel González Urueña, "Descripción de una cura de quemadura causada por la inflamación de los gases que se desprenden de las letrinas," *La Gaceta Médica de México* 1, no. 368 (1836): 368–70.

21. Broadside engraving, José Guadalupe Posada, "Espanto Teremotos y Formidable Incendio en San Francisco California! ¡La Ciudad Entre Las Llamas!—Millares de Victimas!—¡Catastrofe Nunca Vista!" (Mexico City: Anto-

nio Vanegas Arroyo, 1906), Jean Charlot Collection, University of Hawaii at Manoa Library.

22. José Guadalupe Posada, "Terrible y conmovedora, espantosa y aterradora catástrofe," in *José Guadalupe Posada: Ilustrador de la vida mexicana* (Mexico City: Fondo Editorial de la Plástica Mexicana, 1963), 290.

23. Elizabeth Netto Calil Zarur and Charles Muir Lovell, *Art and Faith in Mexico: The Nineteenth-Century Retablo Tradition* (Albuquerque: University of New Mexico Press, 2001).

24. Sherry Fields, *Pestilence and Headcolds: Encountering Illness in Colonial Mexico* (New York: Columbia University Press, 2008).

25. Gloria Fraser Giffords, *Mexican Folk Retablos*, rev. ed. (Albuquerque: University of New Mexico Press, 1992), 150.

26. *Mexico: Splendors of Thirty Centuries* (New York: Metropolitan Museum of Art, 1990), 530–31.

27. Joanna Bourke explains that the rise of disaster experts (people who conducted experiments and determined probabilities of disasters) ultimately helped to prevent large-scale panics. "Fear and Anxiety: Writing about Emotion in Modern History," *History Workshop Journal* 55, no. 1 (2003): 119.

28. Virginia García Acosta and Gerardo Suarez Reyñoso, *Los sismos en la historia de Mexico*, vol. 1 (Mexico City: Universidad Nacional Autónoma de México, 1996), 474–85.

29. M. Nafera to S. Alcalde Municipal, March 23, 1866, AHDF: AGDF, Policía, Incendios, vol. 3649, exp. 59, fs. 23.

30. "Higiene y luz," *La Bandera de Juárez*, 1873.

31. "Hogar bien alumbrado. La luz electrica," *El Diario*, August 17, 1911, 8.

32. "La velada literaria en honor de la Virgen de Guadalupe," *La Voz de México*, October 22, 1895, 2–3.

33. Matthews, *Civilizing Machine*, 143–97.

34. Christopher R. Boyer, *Political Landscapes: Forests, Conservation, and Community in Mexico* (Durham, NC: Duke University Press, 2015), 41–42.

35. For a small sampling of the fires that affected railroad stations or cars, see Diligencia por incendio de una plataforma del ferrocarril de manzanillo, January 23, 1882, AGN: Suprema Corte de Justicia de la Nación [hereafter SCJN] Penal, exp. 139; Averiguación del incendio que sufrió un carro del ferrocarril, October 16, 1885, AGN: SCJN Penal, exp. 308; Averiguación por incendio del puente la reforma del ferrocarril de Morelos, June 30, 1887, AGN: SCJN Penal, exp. 63; "Miscelanea," *Voz de México*, March 19, 1887, 3; *El Siglo Diez y Nueve*, June 7, 1893, 3.

36. "Incendio en San Baltasar Temascalac," *Voz de México*, March 1, 1890, 3.

37. "Un tren de pasajeros se cayo al abismo en el C. de Toluca," *El Imparcial*, August 22, 1911, 3.

38. El Comandante del Cuerpo de Bomberos to the President of the Ayuntamiento, December 30, 1898, AHDF: AGDF, Aguas en General, vol. 41, exp. 504, fs. 2–3.

39. Beatty, *Technology and the Search for Progress*, 177–78.

40. Patrick Frank, *Posada's Broadsheets: Mexican Popular Imagery, 1890–1910* (Albuquerque: University of New Mexico Press, 1998), 186–91.

41. "Un incendio destruya parte del Mercado de Tepito," *El Imparcial*, January 18, 1913, 2; "El incendio del Mercado de Tepito," *El Imparcial*, January 19, 1913, 6.

42. Broadside engraving, José Guadalupe Posada, "Quemazón en el Baratillo de Tepito" (Mexico City: Antonio Vanegas Arroyo, 1913), Jean Charlot Collection, University of Hawaii at Manoa Library.

43. "¡¡¡La Gran Destrucción y Terrible Incendio de la Plaza de Toros de Puebla. El 12 de Enero del Presente Año!!! ¡¡¡Un Muerto, Muchos Heridos y Contusos!!!" in *Posada's Popular Mexican Prints: 273 Cuts by José Guadalupe Posada*, ed. Roberto Berdecio and Stanley Appelbaum (New York: Dover, 1972).

44. Rossana Reguillo, "The Social Construction of Fear: Urban Narratives and Practices," in *Citizens of Fear: Urban Violence in Latin America*, ed. Susana Rotker (New Brunswick, NJ: Rutgers University Press, 2002): 190–93; Lila Abu-Lughod and Catherine A. Lutz, "Introduction: Emotion, Discourse, and Politics of Everyday Life," in *Language and the Politics of Emotion*, ed. Catherine A. Lutz and Lila Abu-Lughod (Cambridge: Cambridge University Press, 1990), 14.

45. Expediente formado con la comunicación en que la Secretaría de Justica excita al Ayuntamiento para que acuerde medidas que eviten los siniestros de incendios entre tanto forma sus reglamentos, March 26, 1885, AGMP, tomo 302, legajo 56, ficha 8454, fs. 83–85.

46. Juan de Zárate, May 21, 1845, AHDF: AGDF, Policía, Incendios, vol. 3649, exp. 47, fs. 1–4.

47. Café Cosmopolita en el que ocurrió un incendio, January 23, 1908, AHDF: AGDF, Fondas y Figones, vol. 1624, exp. 228, fs 4–5.

48. Anton Benjamin Rosenthal, "Spectacle, Fear, and Protest: A Guide to the History of Urban Public Space in Latin America," *Social Science History* 24, no. 1 (2000): 34; Gareth Jones, "The Latin American City as Contested Space: A Manifesto," *Bulletin of Latin American Research* 13, no. 1 (1994): 1–12; Henri Lefebvre, "Reflections on the Politics of Space," trans. Michael J. Enders, *Antipode* 8, no. 2 (1976): 30–37.

49. Reglamentos de incendios para la Ciudad de México, November 29, 1829, AHDF: AGDF, Policía, Incendios, vol. 3649, exp. 41, fs. 1–7.

CHAPTER TWO. SCIENCE OF REGULATION

1. Tenorio-Trillo, *I Speak of the City*, 284.

2. This led officials and hygienists to develop urban services including sewage disposal, potable water, and waste collection, as well as personal hygiene programs such as vaccinations, inoculations, and cleanliness campaigns. See Martin V. Melosi, *The Sanitary City: Urban Infrastructure in America from Colonial Times to the Present* (Baltimore, MD: Johns Hopkins University Press, 2000), 103–5.

3. El virrey Juan Vicente de Guemes Pacheco de Padilla, Reglamento para evitar incendios, September 18, 1790, AHDF: AGDF, Policía, Incendios, vol. 3649, exp. 6, fs. 1–19; El Corregidor don Bernardo Bonabia, petrechos y utensilios contra incendios, April 8, 1790, AHDF: AGDF, Policía, Incendios, vol. 3649, exp. 7, fs. 1–3; Bernardo Bonabia ordena que los dueños de tlapalería no tengan luz ni lumbre, April 29, 1790, AHDF: AGDF, Policía, Incendios, vol. 3648, exp. 4, fs. 1–3; José M. del Castillo Velasco, *Colección de leyes, supremas órdenes, bandos, disposiciones de policía y reglamentos municipals de administración del Distrito Federal*, 2nd ed. (Mexico City: Impreso por Castillo Velasco é Hijos, 1874), 213.

4. Reglamentos de incendios para la Ciudad de México, June 3, 1829, AHDF: AGDF, Policía, Incendios, vol. 3649, exp. 41, fs. 1–7.

5. José María González Mendoza, "Policía en caso de incendio," *Gaceta de los tribunales de la República Mexicana* 3 (September 20, 1862): 818–20.

6. González Mendoza, "Policía en caso de incendio," 818–20.

7. "Inspección general de policia del Distrito Federal," *El Siglo Diez y Nueve*, April 24, 1870, 4.

8. Andrés Lira González, *Comunidades indígenas frente a la Ciudad de México. Tenochtitlán y Tlatelolco, sus pueblos y barrios, 1812–1919* (Zamora, Michoacán: El Colegio de Michoacán–El Colegio de Mexico–Consejo Nacional de Ciencia y Tecnología, 1983).

9. El C. José Maria Castro, Gobernador del Distrito Federal, Disposiciones para evitar incendios, September 26, 1871, AHDF, Gobierno del Distrito Federal, caja 41, exp. 10.

10. Comisión de Policía, March 22, 1870, AHDF: AGDF, Policía, Incendios, vol. 3649, exp. 66.

11. *Boletín Oficial*, February 28, 1905, AHDF: AGDF, Consejo Superior de Gobierno del Distrito: Expendios de Bebidas, Exposiciones, Escuelas, vol. 597, exp. 1.

12. One illustrative example includes a fire that occurred in 1874 at a pulquería located on Tacuba Street; see "Incendio," *El Siglo Diez y Nueve*, December 1, 1874, 6.

13. "Incendio—Por causas que se ignoran," *Voz de México*, July 12, 1896, 3.

14. "Fire Does Damage to the Bookbinding House," *Mexican Herald*, September 26, 1909, 12.

15. Diego Armus and Adrián López Denis, "Disease, Medicine, and Health," in *The Oxford Handbook of Latin American History*, ed. José C. Moya (Oxford: Oxford University Press, 2010), 428; Johan P. Mackenbach, "Politics Is Nothing but Medicine at a Larger Scale: Reflections on Public Health's Biggest Idea," *Journal of Epidemiology and Community Health* 63, no. 3 (2009): 181–84; Owen Roberts, "The Politics of Health and the Origins of Liverpool's Lake Vyrnwy Water Scheme, 1871–92," *Welsh History Review/Cylchgrawn Hanes Cymru* 20, no. 2 (2000): 308–35.

16. Claudia Agostoni, "Discurso médico, cultura higiénica y la mujer en la Ciudad de México al cambio de siglo (XIX–XX)," *Mexican Studies/Estudios Mexicanos* 18, no. 1 (2002): 1–22; Mariola Espinosa, "Globalizing the History of Disease, Medicine, and Public Health in Latin America," *Isis* 104, no. 4 (2013): 800; Steven Palmer, *From Popular Medicine to Medical Populism: Doctors, Healers, and Public Power in Costa Rica, 1800–1940* (Durham, NC: Duke, 2003); Christopher C. Sellers, *Hazards of the Job: From Industrial Disease to Environmental Health* (Chapel Hill: University of North Carolina Press, 1997), 105; Tenorio-Trillo, *I Speak of the City*, 289; Nancy Tomes, "The Private Side of Public Health: Sanitary Science, Domestic Hygiene and the Germ Theory, 1870–1900," *Bulletin of the History of Medicine*, no. 64 (1990): 509–39; Georges Vigarello, *Lo limpio y lo sucio. La higiene del cuerpo desde la edad media* (Madrid: Alianza Editorial, 1985), 240–45.

17. Afonso Soares de Oliveira Sobrinho, "São Paulo e la Ideologia Higienista entre os séculos XIX e XX a utopia de civilidade," *Sociologia* 15, no. 32 (2013): 210–35.

18. Roberto Gayol, "Reflexiones sugeridas por el Art. 257 del Codigo Sanitario que se refiere a las obras públicas que interesan a la higiene," *Boletín del Consejo Superior de Salubridad del Distrito Federal, 3a Época* 3, no. 5 (1897): 139.

19. M. A. Posalagua, *Estudio para la formación de hospitales generales en la Ciudad de México* (Mexico City: Imprenta de Comercio de Nabor Chávez, 1874), 5–6.

20. Dr. Julio Delobel (de Noyon), "Higiene del escolar," *Boletín del Consejo Superior de Salubridad, 3a Época* 7, no. 1 (June 31, 1901): 1–19.

21. S. Morales Pereira, "La más grande de las ensenanzas que deben dares a nuestro pueblo: ¿qué es la higiene? ¿qué fines persigue?" *El Diario*, December 27, 1906, 6.

22. Robert L. Harris, "PCIH Presentation: The Public Health Roots of Industrial Hygiene," *American Industrial Hygiene Association Journal* 58, no. 3 (1997): 176–79.

23. Victor F. Pacheco Salazar and Héctor L. Ocaña Servín, "La legislación ambiental en México," in *Daños a la salud por contaminación atmosférica*, ed. Favio Gerardo Rico Méndez, Rafael López Castañares, and Ezequiel Jaimes Figueroa (Toluca: Universidad Autónoma del Estado de México, 2001), 421.

24. Arthur F. McEvoy, "Working Environments: An Ecological Approach to Industrial Health and Safety," *Technology and Culture* 36, no. 2 (1995): S163.

25. Walter Licht, *Working for the Railroad: The Organization of Work in the Nineteenth Century* (Princeton, NJ: Princeton University Press, 1987), 181.

26. Moisés González Navarro, "Illness and Mortality," in *The Age of Porfirio Díaz, Selected Readings*, ed. Carlos B. Gil (Albuquerque: University of New Mexico Press, 1977), 112.

27. *Código sanitario de los Estados Unidos Mexicanos* (Mexico: Imprenta de la Patria, 1891), 6–7.

28. Gayol, "Reflexiones sugeridas por el Art. 257 del Código Sanitario," 139.

29. Gabriela Soto Laveaga and Claudia Agostoni, "Science and Public Health in the Century of Revolution," in *A Companion to Mexican History and Culture*, ed. William H. Beezley (Malden, MA: Wiley-Blackwell, 2011), 562.

30. Paul Ross, "Mexico's Superior Health Council and the American Public Health Association: The Transnational Archive of Porfirian Public Health, 1887–1910," *Hispanic American Historical Review* 89, no. 4 (2009): 574.

31. "Iniciativa: para la reglamentación en el Distrito Federal de los establecimientos peligrosos, insalubres, e incomodos," *Boletín del Consejo Superior de Salubridad* 3, no. 1–2 (August 31, 1882): 1–3.

32. Artículo 157 del Código Sanitario, November 23, 1911, AHDF: AGDF, Consejo Superior de Gobierno del Distrito: Expendios de Bebidas, Exposiciones, Escuelas, vol. 597, exp. 1.

33. Ross, "Mexico's Superior Health Council," 577.

34. Caracas had similar public health reforms with its 1871 regulations of factories and public buildings, which eventually became city ordinances in 1910. Arturo Almandoz, "The Shaping of Venezuelan Urbanism in the Hygiene Debate of Caracas, 1880–1910," *Urban Studies* 37, no. 11 (2000): 2085.

35. Irving A. Watson, "The Republic of Mexico—Medicine Curative and Preventive," *The Sanitarian, a Monthly Magazine Devoted to the Preservation of Health, Mental and Physical Culture* 29, no. 272 (1892): 122.

36. Mauricio Tenorio-Trillo, *Mexico at the World's Fairs: Crafting a Modern Nation* (Berkeley: University of California Press, 1996), 157.

37. "Un ramal peligroso," *El Diario*, October 28, 1906.

38. *Código sanitario de los Estados Unidos Mexicanos*, 35.

39. "Expediente: Relativo a mejoras introducidas en la fabricación de los

cerillos por los Sres. Lascurain y Compañía," *Boletín del Consejo de Salubridad del Distrito Federal* 3, no. 3–4 (September 30, 1882): 41–42.

40. "Cuarta Comisión de Fábricas e Industrias," *Boletín del Consejo de Salubridad del Distrito Federal* 2, no. 9 (March 31, 1882): 138–39.

41. "Informe: Relativo a las condiciones higiénicas que deben satisfacer los teatros y otras salas de espectáculo," *Boletín del Consejo de Salubridad del Distrito Federal* 3, no. 7–8 (February 28, 1883): 100–101.

42. *Código sanitario de los Estados Unidos Mexicanos*, 42.

43. J. J. R. de Arellano and D. Orvañanos, "Consejo Superior de Salubridad," *El Estudio: Semanario de Ciencias Medicas* 1, no. 27 (December 9, 1889): 431–32.

44. David Rosner and Gerald Markowitz, *Deceit and Denial: The Deadly Politics of Industrial Pollution* (Berkeley: University of California Press, 2002), 1–12.

45. Winthrop Talbot, "Some Economic Aspects of Factory Hygiene," *American Journal of Public Health* 2, no. 10 (1912): 774.

46. Sellers, *Hazards of the Job*, 8. The most famous exposé of the lack of industrial safety and health was Upton Sinclair's *The Jungle* (New York: Doubleday, Jabber & Company, 1906). The infamous Triangle Shirtwaist factory in New York brought safety practices in factories to the forefront of industrial hygiene; see Arthur McEvoy, "The Triangle Shirtwaist Factory Fire of 1911: Social Change, Industrial Accidents, and the Evolution of Common-Sense Causality," *Law and Social Inquiry* 20, no. 2 (1995): 621–51.

47. Mark Aldrich, *Safety First: Technology, Labor, and Business in the Building of American Work Safety, 1870–1939* (Baltimore, MD: Johns Hopkins University Press, 1997), 122–25.

48. Eduardo Licéaga, "Preamble to the Mexican Sanitary Code," in *Historia de la Salubridad y de la Asistencia en México*, ed. José Alvarez Amézquita et al. (Mexico City: Secretaría de Salubridad y Asistencia, 1960), 1:327.

49. A recent study of municipal services in Morelia, Michoacán, discovered a similar shift from community responsibility to government action regarding sewage disposal, public health, and street maintenance. Christina Jiménez, "Popular Organizing for Public Services: Residents Modernize Morelia, Mexico, 1880–1920," *Journal of Urban History* 30, no. 4 (2004): 495–518.

CHAPTER THREE. CONTROLLING THE FLAMES—THE FIRE BRIGADE

1. Amy S. Greenberg, "The Origins of the Municipal Fire Department: Nineteenth-Century Change from an International Perspective," in *Municipal Services and Employees in the Modern City: New Historical Approaches*, ed. Michèle Dagenais, Irene Maver, and Pierre-Yves Saunier (Aldershot, UK: Ashgate,

2003), 47–65; Hubert Lussier, *Les Sapeurs-Pompiers au XIXe siècle: Associations volontaires en milieu populaire* (Paris: L'Harmattan/Association des Ruralistes Français, 1987).

2. "El bombero de Valparaiso," *El Siglo Diez y Nueve*, January 7, 1852, 2.

3. "Boletin del 'Monitor,'" *El Monitor Republicano*, September 28, 1882, 1.

4. Proyecto para formar una sostenida para los fondos Municipales, 1850, AHDF: AGDF, Policía: Incendios, vol. 3649, exp. 51.

5. Don Juan Turín, sobre que se forme un cuerpo de Zapadores Bomberos y se le nombre su Instructor, April 25, 1854, AHDF: AGDF, Policía: Incendios, vol. 3649, exp. 57.

6. Luís de la Barrera, June 18, 1850, AHDF: AGDF, Policía: Incendios, vol. 3649, exp. 51.

7. Orden suprema para que se de una noticia de las bombas que ocurran para apagarlos y del reglamento respectivo, September 22, 1848, AHDF: AGDF, Policía: Incendios, vol. 3649, exp. 50, fs. 2–3.

8. C. Juan J. Baz, Gobernador del Distrito, Decreto expedido por la secretaría de estado y del despacho de Gobernación sobre la organización de un cuerpo de policía denominado de zapadores bomberos, se adjunta el Reglamento para dicho cuerpo de policía, 1856, AGN: GD 127, caja 451, exp. 16. Officials in Tamaulipas made an earlier attempt at recruitment; see "Bases para la organización de una compañía de bomberos en esta ciudad," *El Universal*, June 5, 1854, 1.

9. Reglamento para el Batallon de Zapadores-Bomberos, June 18, 1850, AHDF: AGDF, Policía: Incendios, vol. 3649, exp. 51; Don Juan Turin sobre que se forme un cuerpo de Zapadores Bomberos y se le nombre su Instructor, April 25, 1854, AHDF: AGDF, Policía: Incendios, vol. 3649, exp. 57; Comisaría Central de Policía, March 23, 1866, AHDF: AGDF, Policía: Incendios, vol. 3649, exp. 59, fs. 27–29.

10. Luis G. Zoldivar, *Apendice a la recopilación de leyes del año de 1859 formado por Luis G. Zoldivar* (Mexico City: Imprenta de A. Boix, 1865), 60–65.

11. Ministère du Commerce, de l'Industrie des postes et des télégraphes, *Exposition Universelle Internationale de 1900. Congrès international des sapeurs-pompiers: tenu à Paris le 12 août 1900: procès-verbal sommaire* (Paris: Imprimerie nationale, 1901), 9.

12. *Memoria acerca del estado y adelantos del Excmo. Ayuntamiento de la Habana presentada por el Sr. D. Miguel Díaz Juárez* (Habana: Imprenta la Tipografía á cargo de Manuel Santana, 1897), 145–46.

13. Ismael Valdés Vergara, *El cuerpo de bomberos de Santiago, 1863–1900* (Valparaíso: Babra y Ca., 1900).

14. "Gacetilla," *El Diario Oficial*, March 20, 1871.

15. E. Burton Steward, *A Fire Department Training School* (New York: New York Underwriters Agency, 1896), n.p.

16. Sobre que sede una gratificación de 10 pesos a los primeros individuos que concurran con las bombas de incendio a donde estos se verifiquen, April 11, 1871, AHDF: AGDF, Policía, Incendios, vol. 3649, exp. 71, fs. 1–5.

17. Even though the professional fire brigade in Mexico City did not appear until 1888, it was one of the first of its kind in Latin America. Panama established a fire department in 1937 (see Carlos Rangel M., *Historia del cuerpo de bomberos de Panamá* [Panamá: Imprenta Nacional, 1962], 11–24), and Tijuana had a brigade in 1922 (see Samuel Meléndez Marín, *Tijuana crece al calor de las llamas* [Tijuana: Editorial Zenit, 1983], 16). In 1825, Guayaquil, Ecuador, established the first brigade in Latin America as a branch of the military, see Modesto Chávez Franco, *Historia general del cuerpo de bomberos de Guayaquil*, 2nd ed. (Guayaquil: Banco Central del Ecuador, 1985), xi.

18. Beezley, *Judas at the Jockey Club*, 10; Manuel Dublán and José María Lozano, *Legislación Mexicana: Colección completa de las disposiciones legislativas expedidas desde la independencia de la República*, vol. 12 (Mexico City: Imprenta y Litografía de Eduardo Dublan y Comp., 1886), 435.

19. El C. Lic. Luís C. Curiel, Gobernador del Distrito Federal, Reglamento de comisiones de policías, inspectors de cuartel, inspectors de manzana, ayudantes de acera y gendarmes bomberos, February 10, 1878, AHDF, Gobierno del Distrito Federal, caja 48, exp. 15, fs. 1–18.

20. Pablo Piccato, *City of Suspects: Crime in Mexico City, 1900–1931* (Durham, NC: Duke University Press, 2001), 13.

21. Paul J. Vanderwood, *Disorder and Progress: Bandits, Police, and Mexican Development* (Wilmington, DE: Scholarly Resources, 1992), 54.

22. Amy S. Greenberg, *Cause for Alarm: The Volunteer Fire Department in the Nineteenth-Century City* (Princeton, NJ: Princeton University Press, 1998), 9–27.

23. Steward, *Fire Department Training School*, n.p.

24. John Kenlon, *Fires and Fire-Fighters: A History of Modern Fire-Fighting with a Review of Its Development from Earliest Times* (New York: George H. Doran Company, 1913), 48.

25. El C. Lic. Luís C. Curiel, Gobernador del Distrito Federal, Reglamento de comisiones de policías, inspectors de cuartel, inspectors de manzana, ayudantes de acera y gendarmes bomberos, February 10, 1878, AHDF, Gobierno del Distrito Federal, caja 48, exp. 15, fs. 15; "Key Plan of City of Mexico," *Sanborn Fire Insurance Maps*, 1905, Perry-Castañeda Library Map Collection, University of Texas Libraries; John Lear, *Workers, Neighbors, and Citizens: The Revolution in Mexico City* (Lincoln: University of Nebraska Press, 2001), 42.

26. "El Bombero," *El Monitor Republicano*, December 9, 1883, 2.

27. *Voz de México*, March 12, 1887, 3; "Buena acción," *Voz de México*, March 15, 1887, 3.

28. "Incendio," *El Siglo Diez y Nueve*, May 16, 1861, 3.

29. Michael Adas, *Machines as the Measure of Men: Science, Technology, and Ideologies of Western Dominance* (Ithaca, NY: Cornell University Press, 1989), 5.

30. Relativo a la compra de una bomba para su extinción, April 6, 1842, AHDF: AGDF, Policía, Incendios, vol. 3649, exp. 43, fs. 22–23.

31. Bomba de Incendio de Repsold, 1845, AHDF: AGDF, Policía, Incendios, vol. 3649, exp. 45.

32. More than half of this amount was used to import pumps from the United States; see José María Garmendia, *República Mexicana, Secretaría de Estado y del Despacho de Hacienda y Crédito Pública: Noticia de la importación y exportación de mercancías en los años fiscales de 1872 á 1873, 1873 á 1874 y 1874 á 1875* (Mexico City: Tipografía de Gonzalo A. Esteva, 1880).

33. Bunker, *Creating Mexican Consumer Culture*, 270–77.

34. "Formidable incendio, el cajón de la Valenciana destruido por el fuego," *El Imparcial*, September 28, 1900, 1.

35. "Key Plan of City of Mexico, Mexico," *Sanborn Fire Insurance Maps*, 1905, Perry-Castañeda Library Map Collection, University of Texas Libraries.

36. "Notables pruebas de incendio," *El Imparcial*, June 5, 1906, 3.

37. "Señor Díaz Used a Fire Hose," *Kansas City Star*, April 6, 1905, 14.

38. "El bombero que resulto herido en el incendio de dia seis, vino al suelo desde una gran altura," *El Imparcial*, January 10, 1909, 5.

39. Pablo Escandon, Glorioso Aniversario del 15 de Septiembre de 1810, September 14, 1888, AHDF, Carteles e Ilustraciones, Festividades, vol. 1070, exp. 120, caja 1, carpeta 58-2.

40. Stephen Neufeld, "Servants of the Nation: The Military in the Making of Modern Mexico, 1876–1911" (PhD diss., University of Arizona, 2009), 245–46.

41. Federico Gamboa, *Santa: A Novel of Mexico City*, trans. John Charles Chasteen (Chapel Hill: University of North Carolina Press, 2012), 66.

42. Cuerpo Auxiliar de Bomberos Voluntarios de Puebla piden uniformes, April 16, 1909, and February 3, 1909, Archivo General Municipal de Puebla [hereafter AGMP], tomo 494, ficha 16945, legajo 21, letra A, fs. 14–79.

43. "Los Bomberos de Veracruz y el Señor Frago," *Voz de México*, March 18, 1887, 2–3.

44. "La velada literaria en honor de la Virgen de Guadalupe," *La Voz de México*, October 22, 1895, 2–3.

45. "La conflagración en Puebla," *El Dictamen*, February 4, 1909.

46. Broadside engraving, José Guadalupe Posada, "Quemazón en el Baratillo de Tepito" (Mexico City: Antonio Vanegas Arroyo, 1913), Jean Charlot Collection, University of Hawaii at Manoa Library.

47. "El incendio de 'El Vulcano' un bombero en la cárcel," *El Imparcial*, June 25, 1906, 6.

48. "Bombero insubordinado dado de baja," *El País*, July 24, 1909, 2; AGMP, tomo 494, ficha 16945, legajo 21, June 1909, fs. 52; AGMP, tomo 499, ficha 17090, exp. 1, letra K, 1910.

49. Bird, *Protection against Fire*, 1–2.

50. "Corrido la quemazón," *Posada's Popular Mexican Prints*, 23.

51. *Overland Monthly: The Awakening of Mexico Centenary of the Republic* 56 (July 1910): 117.

52. "Don Bromista. Humorous Features of the Sewer Work in the Streets of Mexico. A Timid Husband and a Brave Wife," *Mexican Herald*, Sunday, June 11, 1899.

53. *Overland Monthly*, 117.

54. Bill Mallon, *The 1900 Olympic Games: Results for All Competitors in All Events, with Commentary* (Jefferson, NC: McFarland and Company, 1998), 249.

55. Ministère du Commerce, de L'Industrie des postes et des télégraphes.

56. Ministère du Commerce, de L'Industrie des postes et des télégraphes, 9.

CHAPTER FOUR. ENGINEERING SAFETY

1. Manuel Rivera Cambas, *México pintoresco, artístico y monumental*, vol. 1 (Mexico City, Editorial Valle de México, 1972), 154–55.

2. José Guadalupe Victoria, "Noticias sobre la antigua plaza y el mercado del volador de la Ciudad de México," *Anales del Instituto de Investigaciones Estéticas* 16, no. 62 (1991): 80–81; "Inspección general de policia del Distrito Federal," *El Siglo Diez y Nueve*, April 24, 1870, 4.

3. García Brito, March 23, 1870, AHDF: AGDF, Rastros y Mercados, vol. 3733, exp. 518, fs. 6–9.

4. Información gubernativa sobre las causas que ocasionaron el incendio del Mercado del Volador, May 3, 1870, AHDF: AGDF, Rastros y Mercados, vol. 3733, exp. 518, fs. 1–3.

5. Pedro Meneses to Regidores, April 23, 1870, AHDF: AGDF, Rastros y Mercados, vol. 3733, exp. 518, fs. 45; Vicente Salazar to Regidores, March 23, 1870, AHDF: AGDF, Rastros y Mercados, vol. 3733, exp. 518, fs. 29; Julio Peña to Regidores, March 25, 1870; AHDF: AGDF, Rastros y Mercados, vol. 3733, exp. 518, fs. 31–32.

6. Gregorio Chavarria to Regidores, March 23, 1870, AHDF: AGDF, Rastros y Mercados, vol. 3733, exp. 518, fs. 28.

7. Joaquin Zerudyar, Resguardo Nocturnal de México to CC Regidores, March 28, 1870, AHDF: AGDF, Rastros y Mercados, vol. 3733, exp. 518, fs. 25–26.

8. Vicente Montes de Roa to Regidores, April 9, 1870, AHDF: AGDF, Rastros y Mercados, vol. 3733, exp. 518, fs. 44.

9. Joaquin Zerudar, Resguardo Nocturnal de México to CC Regidores, March 28, 1870, AHDF: AGDF, Rastros y Mercados, vol. 3733, exp. 518, fs. 25–26.

10. Cayetano Gómez Pérez to C. Regidores, March 25, 1870, AHDF: AGDF, Rastros y Mercados, vol. 3733, exp. 518, fs. 13–16.

11. Julio Peña to Regidores, March 25, 1870, AHDF: AGDF, Rastros y Mercados, vol. 3733, exp. 518, fs. 31–32; Vicente Garces, Guarda del Mercado to Regidores, March 25, 1870, AHDF: AGDF, Rastros y Mercados, vol. 3733, exp. 518, fs. 32–33; Vicente Constantino, cobrador de las puertas de la plaza, to Regidores, March 25, 1870, AHDF: AGDF, Rastros y Mercados, vol. 3733, exp. 518, fs. 39; José Margarito Vazquez to Regidores, March 25, 1870, AHDF: AGDF, Rastros y Mercados, vol. 3733, exp. 518, fs. 41; Victinano Chávez to Regidores, March 25, 1870, AHDF: AGDF, Rastros y Mercados, vol. 3733, exp. 518, fs. 43.

12. Cayetano Gómez Pérez to C. Regidores, March 25, 1870, AHDF: AGDF, Rastros y Mercados, vol. 3733, exp. 518, fs. 13–16.

13. Manuel Patiño to Regidores, March 25, 1870, AHDF: AGDF, Rastros y Mercados, vol. 3733, exp. 518, fs. 22–23.

14. "Mercados," *El Siglo Diez y Nueve*, March 22, 1870, 3.

15. "Peligro," *El Siglo Diez y Nueve*, May 14, 1872, 3.

16. "Expediente Relativo a la visita de la Plaza de Mercado," *Boletín del Consejo Superior de Salubridad del Distrito Federal* 1, no. 3 (September 20, 1880): 26.

17. Belem Clark de Lara and Mariana Flores Monroy, *El renacimiento: periódico literario, segunda época* (Mexico City: Universidad Autónoma de México, 2006), 192.

18. Wiebe E. Bijker, Thomas P. Hughes, and Trevor Pinch, *The Social Construction of Technological Systems: New Directions in the Sociology and History of Technology* (Cambridge, MA: MIT Press, 1989), 10.

19. A critique of the social construction of science is that it presupposes that historical actors are oblivious to the ways in which belief systems, political views, and economic situations manipulate their overall patterns of thought; see David Demeritt, "Ecology, Objectivity, and Critique in Writings on Nature and Human Societies," *Journal of Historical Geography* 20, no. 1 (1994): 27.

20. Guerra, *México*, 380; Luis Sánchez Agesta, *El pensamiento politico del despotismo ilustrado* (Madrid: Instituto de Estudios Políticos, 1953), 16.

21. This trend was common throughout Latin America; see Angela Maria Alonso, *Idéias em movimento: a geração 1870 na crise do Brasil-Império* (São Paulo: ANPOCS, Paz e Terra, 2002); Lilia Moritz Schwarcz, *The Spectacle of the Races: Scientists, Institutions, and the Race Question in Brazil*, trans. Leland Guyer (New York: Hill and Wang, 1999); Marta de la Vega, *Evolucionismo versus positivismo: estudio teórico sobre el positivismo y su significación en América Latina* (Caracas, Venezuela: Monte Avila Editores Latinoamericana, 1998).

22. Jorge L. Tamayo, *Breve reseña sobre la escuela nacional de ingeniera* (Mexico City: Imprenta La Esfera, 1958), 53.

23. José Ramón de Ibarrola, *Apuntes sobre el desarrollo de la ingeniería en México y la educación del ingeniero* (Mexico City: Tipografía de la viuda de F. Díaz de León, 1911), 6.

24. Ministro de Justicia e Instrucción Pública, "Prácticas de Alumnos," July 7, 1870, Universidad Nacional Autónoma de México [hereafter UNAM], Fondo Escuela Nacional de Ingeniería (ENI), Asuntos Escolares Subramo, caja 30, exp. 11, fs. 4–5; Roberto Gayol, Plano de la Ciudad de México, indica el trayecto que deben seguir los tubos de distribución, colectores y atarjeas laterales, 1898, AHDF, Planoteca, Gaveta 1, Plano: 8.

25. Robert H. Kargon and Scott G. Knowles, "Knowledge for Use: Science, Higher Learning, and America's New Industrial Heartland, 1880–1915," *Annals of Science* 59, no. 1 (2002): 14.

26. Ramón de Ibarrola, *Apuntes sobre el desarrollo*, 6.

27. Ramón de Ibarrola, *Apuntes sobre el desarrollo*, 23.

28. Francisco de Garay, *Discurso pronunciado por el ingeniero Francisco de Garay en la Asociación de Ingenieros Civiles y Arquitectos, al tomar posesión de la presidencia de la misma* (Mexico City: Imprenta del Comercio de Dublán y Chávez, 1877), 8.

29. Andrew Abbott, *The System of Professions: An Essay on the Division of Expert Labor* (Chicago: University of Chicago Press, 1988); Monte A. Calvert, *The Mechanical Engineer in America, 1830–1910: Professional Cultures in Conflict* (Baltimore, MD: Johns Hopkins University Press, 1967), 191; Bruce Sinclair, *A Centennial History of the American Society of Mechanical Engineers, 1880–1980* (Toronto: University of Toronto Press, 1980), 25–26 and 61; Amy Slaton, *Reinforced Concrete and the Modernization of American Building, 1900–1930* (Baltimore, MD: Johns Hopkins University Press, 2001), 21–25.

30. Jorge E. Hardoy, "Theory and Practice of Urban Planning in Europe, 1850–1930," in *Rethinking the Latin American City*, ed. Richard M. Morse and Jorge E. Hardoy (Washington, DC: Woodrow Wilson Center Press), 23.

31. Tamayo, *Breve reseña sobre la escuela*, 46.

32. Kargon and Knowles, "Knowledge for Use," 14.

33. Dublán and Lozano, *Legislación Mexicana*, 434.

34. Guillermo N. Perja to Secretario del Ayuntamiento, October 29, 1900, AGDF: AHDF, Oficinas en General, vol. 3378, exp. 373, fs. 1–7.

35. Reglamento para el Batallón de Zapadores-Bomberos, 1850, AHDF: AGDF, Policía: incendios, vol. 3649, exp. 51.

36. Plan y presupuesto de una compañía de zapadores bomberos y sus reglamentos según el estilo Francia, April 25, 1854, AHDF: AGDF, Policía: Incendios, vol. 3649, exp. 57.

37. For a sampling of court cases that requested fire inspections, see AHDF: AGDF, Justicia Negocios Judiciales, 1894–1916, vol. 2727, exps. 168, 234, 292, 362; Bruce L. R. Smith, *The Advisers: Scientists in the Policy Process* (Washington, DC: Brookings Institution, 1992), 16.

38. Luís C. Curiel, relativo a la concurrencia de los ingenieros de la ciudad y del director de aguas a los incendios, 1878, AHDF: AGDF, Policía, Incendios, vol. 3649, exp. 80.

39. Los vecinos de la 2a calle de Tulipán, manifiestan temor de que ocurra un incendio en la fábrica de cerillos La Fortuna, September 2, 1904, AHDF: AGDF, Gobierno del Distrito, Fábricas, vol. 1602, exp. 151, fs. 1–6.

40. Se proceda a remover los escombros provenientes del incendio en el edificio del Palacio de Hierro, May 21, 1914, AHDF: AGDF, Consejo Superior de Gobierno del Distrito: Festividades (Miscelánea), vol. 603, exp. 1.

41. C. John Hexamer, *On the Prevention of Fires in Theatres* (Philadelphia: Merrihew Print, 1882), 1.

42. Formado con todo lo relativo al cuerpo municipal de bomberos, February 3, 1909, AGMP, tomo 494, ficha 16945, legajo 21, f. 14–79; "El incendio en el Teatro Principal de Puebla," *El Imparcial*, July 30, 1902, 1; "El incendio del Teatro Principal," *El Imparcial*, August 3, 1902, 1; "El incendio del teatro de la ópera en Paris," *El Siglo Diez y Nueve*, December 10, 1873, 3.

43. August Foelsch, *Theaterbrände und die zur Verhütung derselben erforderlichen Schutz-Massregelu; mit einem Verzeichniss von 523 abgebrannten Theatern* (Hamburg: O. Meissner, 1878).

44. Édouard Choquet, *Les Incendies dans les Théâtres* (Paris: Baudry, 1886); William Paul Gerhard, *Theatre Fires and Panics: Their Causes and Prevention* (New York: John Wiley and Sons, 1896), 12–19; Sara E. Wermiel, *The Fireproof Building: Technology and Public Safety in the Nineteenth-Century American City* (Baltimore, MD: Johns Hopkins University Press, 2000), 198–88.

45. Hexamer, *On the Prevention of Fires in Theatres*, 13.

46. Gerhard, *Theatre Fires and Panics*, 27.

47. Concejal Ing. Abraham Chávez to Secretario del Consejo Superior de Gobierno, August 7, 1912, AHDF: AGDF, Consejo Superior de Gobierno del distrito, diversiones públicas, vol. 596, exp. 19.

48. Juan Félipe Leal, Eduardo Barraza, and Carlos Flores, *El arcón de las vistas: cartelera del cine en México, 1896–1910* (Mexico City: Universidad Nacional Autónoma de México, 1994), 44–46.

49. Leo Enticknap, "The Film Industry's Conversion from Nitrate to Safety Film in the Late 1940s: A Discussion of the Reasons and Consequences," in *This Film Is Dangerous: A Celebration of Nitrate Film*, ed. Roger Smither and Catherine Surowiec (Brussels: FIAF, 2002), 209.

50. Heather Heckman, "Burn after Viewing, or, Fire in the Vaults: Nitrate Decomposition and Combustibility," *American Archivist* 73, no. 2 (2010): 483–506.

51. "Theater Fire in a Mexican Western Port," *Hopkinsville Kentuckian* 30, no. 21, February 18, 1909; "Todavía las víctimas de Acapulco," *El País*, March 7, 1909, 1; "Para las víctimas de Acapulco," *El País*, March 14, 1909, 1.

52. Juan Palacios, March 22, 1870, AHDF: AGDF, Policía, Incendios, vol. 3649, exp. 60; Circular a los dueños de teatros previniéndoles que conforme al Reglamento vigente, tengan siempre disponible una bomba para incendios, 1870, AHDF: AGDF, Policía, Incendios, vol. 3649, exp. 62.

53. Comisión de diversiones públicas to Presidente del Ayuntamiento Constitucional de México, April 13, 1870, AHDF: AGDF, Policía, Incendios, vol. 3649, exp. 64.

54. Sobre que los teatros de la ciudad tengan sus respectivas bombas, AHDF: AGDF, Policía, Incendios, vol. 3649, exp. 67, fs. 3.

55. Ángel González de la Torre to Secretario del Gobierno del Distrito, July 25, 1878, AHDF: AGDF, Gobierno del Distrito, Obras Públicas, vol. 1755, exp. 287.

56. Manuel Bandera to Secretario del Gobierno del Distrito, August 17, 1878, AHDF: AGDF, Gobierno del Distrito, Obras Públicas, vol. 1755, exp. 287.

57. Juan N. Cortina to Gobernador del Distrito Federal, September 4, 1878, AHDF: AGDF, Gobierno del Distrito, Obras Públicas, vol. 1755, exp. 287.

58. Incendio del Teatro llamado de la Zarzuela en la tarde del 2 de noviembre del presente año, 1874, AHDF: AGDF, Teatros, vol. 4017, exp. 79.

59. Juan Bribiesca, Secretario, Reglamento para prevenir los incendios en los teatros, May 6, 1888, AHDF: AGDF, caja 58, exp. 25.

60. "Abraham Chávez," *El País*, July 19, 1905, 2.

61. Concejal Ingeniero Abraham Chávez to Secretario del Consejo Superior de Gobierno del Distrito, February 7, 1912, AHDF: AGDF, Diversiones Públicas, vol. 807, exp. 1323.

62. Juan Félipe Leal and Eduardo Barraza, "Inicios de la reglamentación cinematográfica en la Ciudad de México," *Revista Mexicana de Ciencias Políticas y Sociales, México*, no. 150 (October–December 1992): 139–77.

63. Proyecto de Reglamento para cinematógrafos, March 7, 1908, AHDF: AGDF, Consejo Superior de Gobierno del Distrito, Expendios de Bebidas, Exposiciones, Escuelas, vol. 597, exp. 1.

64. *El Imparcial*, September 5, 1902, 4; Julio L. Perié to Abraham Chávez, March 1, 1912, AHDF: AGDF, Diversiones Públicas, vol. 807, exp. 1323; Juan Bribiesca, *Memoria documentada de los trabajos municipales de 1900*, vol. 1 (Mexico City: La Europea, 1901), 523. In her assessment of theater life in Mexico City, Fanny Calderón de la Barca noted that the whole theater reeked of smoke because audience members, attendants, and even performers smoked constantly—see *Life in Mexico: The Letters of Fanny Calderón de la Barca with New Material from the Author's Private Journals*, ed. Howard T. Fisher and Marion Hall Fisher (Garden City, NY: Doubleday, 1966), 114.

65. Consejo Superior de Salubridad to Comisión de Teatros y Inspectores de Diversiones Públicas del Gobierno del Distrito, 1912, AHDF: AGDF, vol. 596, exp. 22.

66. Adolfo Díaz Rugama, "Distribución y legislación de aguas en las ciudades," in *Concurso Científico. Asociación de Ingenieros y Arquitectos. Discurso pronunciado en la sesión del 22 de julio de 1895* (Mexico City: Oficina de Tipografía de la Secretaría de Fomento, 1895).

67. Ramón de Ibarrola, *Apuntes sobre el desarrollo de la ingeniería en México y la educación del ingeniero*, 6.

68. Institution of Civil Engineers, *Power for the Use of Man: A Series of Addresses to the General Assembly of the Institution of Civil Engineers, Commemorating the 150th Anniversary of the Royal Charter of the Institution granted on 3 June 1828* (London: Institution of Civil Engineers, 1978).

69. Adas, *Machines as the Measure of Men*, 3.

70. Gilberto Crespo Martínez, *Las patentes de invención* (Mexico City: Oficina Tipografía de la Secretaría de Fomento, 1897): 9–11, 85.

71. Paul R. Josephson, *Industrialized Nature: Brute Force Technology and the Transformation of the Natural World* (Washington, DC: Island Press/Shearwater Books, 2002), 3.

72. Manuel Perló Cohen, *El paradigma porfiriano: historia del desagüe del valle de México* (Mexico City: Universidad Autónoma de México, 1999).

73. Jesús Galindo y Villa, *Reseña histórica-descriptiva de la Ciudad de México que escribe el Regidor del Ayuntamiento, por encargo del Señor Presidente de la misma corporación D. Guillermo Landa y Escandón, y expresamente para los delegados a la segunda conferencia internacional americana* (Mexico City: Díaz de Leon, 1901), 166–67.

74. Inspector General de Policia to Gobierno del Distrito Federal, May 16, 1872, AHDF: AGDF, Aguas en General, vol. 36, exp 202, fs. 1–2.

75. Se ordena al Ayuntamiento de la capital, remita a este Gobierno un plano de todas las cañerías de agua que cruzan las calles de la capital, para tomarla en caso de incendio, 1870–1877, AHDF: AGDF, Goberino del Distrito, Aguas, vol. 1311, exp. 342.

76. Carlos A. de Medina, *Exposición que hace el Ingeniero Carlos A. de Medina a todos los habitantes de la Ciudad de México sobre las grandes ventajas que trae consigo para la capital* (Mexico City: Imprenta y Litografía de Dublan y Ca., 1884), 18.

77. Luís Salazar, "On the Distribution of Water in the City of Mexico," *International Engineering Congress* 30 (1893): 344.

78. Soil liquefaction caused extensive property damage and loss of life during the 1985 earthquake in Mexico City; see Elena Poniatowska, *Nothing, Nobody: The Voices of the Mexico City Earthquake* (Philadelphia, PA: Temple University Press, 1995).

79. Salazar, "On the Distribution of Water," 348.

80. Matthew Vitz, "Revolutionary Environments: The Politics of Nature and Space in the Valley of Mexico, 1890s–1940s" (PhD diss., New York University, 2010), 38.

81. Salazar, "On the Distribution of Water," 344–47.

82. "Key Plan of City of Mexico, Mexico," *Sanborn Fire Insurance Maps*, 1905, Perry-Castañeda Library Map Collection, University of Texas Libraries.

83. Comandante del Cuerpo de Bomberos to Presidente del Ayuntamiento, March 8, 1897, AHDF: AGDF, Aguas en General, vol. 40, exp. 467, fs. 5–8.

84. Juan Bribiesca, *Memoria documentada de los trabajos municipales de 1900*, vol. 1 (Mexico City: La Europea, 1901), 343.

85. Luis G. de Ansorena, May 10, 1865, AHDF: AGDF, Policía, Incendios, vol. 3649, exp. 58.

86. Inspector General to Secretario del Gobierno del Distrito, April 15, 1903, AHDF: AGDF, Gobierno del Distrito, Aguas, vol. 1327, exp. 1456.

87. El Comisionario de la 6a demarcación remite el acta levantada con motivo que Andrés López y otros sacaban agua de la toma para incendios en Chapultepec y San Antonio, 1904–1907, AHDF: AGDF, Gobierno del Distrito, Aguas, vol. 1328, exp. 1487; La dirección general de obras públicas pide se prevenga a la policía cuide del público no haga uso de la válvulas de incendio, 1904–1907, AHDF: AGDF, Gobierno del Distrito, Aguas, vol. 1328, exp. 1517.

88. Vitz, "Revolutionary Environments," 49.

89. Jefe de Bomberos to Administrador de Paseos, April 28, 1896, AHDF: AGDF, Aguas en General, vol. 40, exp. 475, fs. 1–2.

90. "Esta ardiendo el petroleo del pozo de 'Dos Bocas': Espantosa Conflagración," *El Imparcial*, March 7, 1908, 7.

91. Juan D. Villarello, "El pozo de petróleo de 'Dos Bocas,'" *Boletín del Instituto Geológico de México* 3 (1909): 17–21.

92. "Llegan los zapadores que fueron á combatir el incendio de Dos Bocas," *El Dictamen*, November 8, 1908, 3.

93. "Como podra apagarse el incendio de 'Dos Bocas': La opinión de un ingeniero americano," *El Imparcial*, July 20, 1908, 3.

94. Juan Palacios, "Memoria sobre incendio del pozo de petroleo de 'Dos Bocas,'" *Boletín de la sociedad Mexicana de Geografia y Estadistica, quinta época* 3 (1908): 23.

95. Minutas y oficios de Luís Salazar, director de la Escuela Nacional de Ingeniería, July 7, 1908, Archivo Histórico de la UNAM, Fondo ENI, Asuntos Escolares Subramo, caja 30. exp 11, fs. 38–54.

96. On-site and laboratory research had become essential to the engineering profession in Mexico and the United States; Kargon and Knowles, "Knowledge for Use," 7.

97. Feliciano J. Garcia Aguirre Mendez and Emilia Valdez, "Dos Bocas: una contribucion a la historia de los desastres en Veracruz," *Instituto de Investigaciones Historico-Sociales* (1995): 105–21.

98. Critics at the time scoffed at this idea, but today most oil wells are capped while they burn.

99. "El incendio del pozo 'Dos Bocas': Doscientas toneladas de grava mezclada con agua, se arrojan cada dia," *El Imparcial*, August 17, 1908, 1 and 8; "El incendio del pozo de petróleo de Dos Bocas," *El Dictamen*, November 11, 1908, 2.

100. "El incendio de pozo de 'Dos Bocas,' *El Imparcial*, August 14, 1908, 4.

101. Glen D. Kuecker, "'The Greatest and the Worst': Dominant and Subaltern Memories of the Dos Bocas Well Fire of 1908," in *The Memory of Catastrophe*, ed. Peter Gray and Kendrick Oliver (Manchester: Manchester University Press, 2004): 65–78.

102. "Como podra apagarse el incendio de 'Dos Bocas,'" *El Imparcial*, July 20, 1908, 3.

103. Juan Palacios, "Memoria sobre incendio del pozo de petroleo de 'Dos Bocas,'" 33.

104. Myrna Santiago, "Rejecting Progress in Paradise: Huastecs, the Environment, and the Oil Industry in Veracruz, Mexico, 1900–1935," *Environmental History* 3, no. 2 (1998): 177.

105. "El incendio de petróleo," *El Pais* (August 1, 1908): 2; Myrna Santiago, *The Ecology of Oil: Environment, Labor, and the Mexican Revolution, 1900–1938* (Cambridge: Cambridge University Press, 2006), 134.

106. Gamboa, *Santa*, 128.

CHAPTER FIVE. INVENTING PROTECTION

1. Gustavo G. Garza Merodio, "Technological Innovation and the Expansion of Mexico City, 1870–1920," *Journal of Latin American Geography* 5, no. 2 (2006): 114–15.

2. Naomi Klein, *The Shock Doctrine: The Rise of Disaster Capitalism* (New York: Metropolitan Books, 2007); Mark Carey, *In the Shadow of Melting Glaciers: Climate Change and Andean Society* (Oxford: Oxford University Press, 2010), 12; Susan Stonich, "International Tourism and Disaster Capitalism: The Case of Hurricane Mitch in Honduras," in *Capitalizing on Catastrophe: Neoliberal Strategies in Disaster Reconstruction*, ed. Nandini Gunewardena and Mark Schuller (Toronto: AltaMira Press, 2008), 47–68.

3. Ernesto Aréchiga Córdoba, "La ciudad y Tepito: a la zaga de la higiene pública," in *Tepito: del antiguo barrio de indios al arrabal*, ed. Ernesto Aréchiga Córdoba (México: UNÍOS, 2003), 171–207; Thomas Benjamin and Marcial Ocasio-Meléndez, "Organizing the Memory of Modern Mexico: Porfirian Historiography in Perspective, 1880s–1980s," *Hispanic American Historical Review* 64, no. 2 (1984): 363; Edward Beatty, "Approaches to Technology Transfer in History and the Case of Nineteenth-Century Mexico," *Comparative Technology Transfer and Society* 1, no. 2 (2003): 167–97.

4. Mario Cerutti, *Empresarios españoles y sociedad capitalista en México (1840–1920)* (Asturias: Fundación Archivo de Indianos, 1995); Warren Dean, *The Industrialization of São Paulo, 1880–1945* (Austin: University of Texas Press, 1969); Peter Evans, *Dependent Development: The Alliance of Multinational, State and Local Capital in Brazil* (Princeton, NJ: Princeton University Press, 1979); Wilson Suzigan, *Industria brasileira: Origem e desenvolvimento* (São Paulo: Editoria Brasiliense, 1986).

5. Nathan Rosenberg, "The International Transfer of Technology: Implications for the Industrialized Countries," in *Inside the Black Box: Technology and Economics*, ed. Nathan Rosenberg (Cambridge: Cambridge University Press, 1982), 245–46.

6. Christopher Freeman, *Technology Policy and Economic Performance: Lessons from Japan* (London: Pinter, 1987); Richard R. Nelson, *National Innovation Systems: A Comparative Analysis* (Oxford: Oxford University Press, 1993); David Hounshell, "Rethinking the History of 'American Technology,'" in *In Context: History and the History of Technology, Essays in Honor of Melvin Kranzberg*, ed. Stephen H. Cutcliffe and Robert C. Post (Bethlehem, PA: Lehigh University Press, 1989), 216–29.

7. *Memoria de la Exposición Municipal de 1874* (México: Imprenta de Díaz de Leon y White, 1874), 29–30.

8. *Recopilación de leyes, decretos y providencias de los poderes legislativo y ejecutivo de la union*, vol. 18 (1874), 560.

9. *Memoria de la Exposición Municipal de 1874*, 34–46.

10. International Municipal Congress and Exposition. Planned for Chicago, Illinois U.S.A., July 14, 1911, AHDF: AGDF, Consejo Superior de Gobierno del Distrito, Exposiciones, vol. 597.

11. Nota dirigida por el Sr. Embajador de los Estados Unidos, June 20, 1911, AHDF: AGDF, Consejo Superior de Gobierno del Distrito, Exposiciones, vol. 597.

12. *El Mensajero*, April 4, 1880, 3; Jesús Galindo y Villa, "Material eléctrico para bomberos," *Boletín Municipal: Órgano Oficial del Ayuntamiento de México*, vol. 1, Mexico, Friday, August 30, 1901, no. 49.

13. *Cleveland Daily*, January 23, 1884; *Independent Utica*, January 18, 1884; *Cincinnati Democrat*, January 10, 1884; *Lowell Daily American*, January 4, 1884; *Orient Daily*, MA, January 17, 1884; "Opinión de los Jefes de Ingenieros de Diferentes Cuerpos de Bomberos, Fall River," January 29, 1884; "Oficinas del Jefe de Ingenieros Departamento de Bomberos de Plainfield," January 16, 1884; AGN: Patentes y Marcas, caja 25, exp. 1146.

14. Carlos Rivas to Secretaría de Gobernación, June 28, 1884, AHDF: AGDF, Policía, Incendios, 3649, exp. 84.

15. Charles M. Martin to Presidente del Ayuntamiento, August 12, 1884, AHDF: AGDF, Policía, Incendios, vol. 3649, exp. 83.

16. Jefe de Bomberos to Gobierno del Distrito, July 12, 1886, AHDF: AGDF, Policía, Incendios, vol. 3649, exp. 85.

17. One notable exception occurred in 1888 when the chief of the fire department attempted a similar publicity stunt. He soaked three wooden tables in petroleum and lit them on fire in the middle of a public plaza. He wanted to show how well the newest fire extinguishers worked, thus assuring the public that the fire brigade was skilled and had the appropriate technology to protect the city. See "El Nuevo aparato," *La Voz de México*, February 22, 1890, 3.

18. "Simulacro de Incendio," *El Imparcial*, July 25, 1902, 2; "Pruebas contra incendio," *El Imparcial*, August 4, 1902, 2.

19. "Pruebas contra incendio," *El Imparcial*, August 5, 1902, 2.

20. "Las granadas 'Harden' contra incendio: un invento maravilloso," *El Mundo Ilustrado*, August 3, 1902.

21. Galindo y Villa, "Material eléctrico para bomberos," 49.

22. Alan L. Durham, *Patent Law Essentials: A Concise Guide*, 2nd ed. (Westport, CT: Praeger, 2004), 1–2.

23. To date, Edward Beatty has provided the most complete analysis of patents and the patenting process in Mexico. See Beatty, *Technology and the*

Search for Progress; Edward Beatty, "Patents and Technological Change in Late Industrialization: Nineteenth-Century Mexico in Comparative Context," *History of Technology* 24 (2002), 130.

24. Juan de la Torre, *Legislación de patentes y marcas* (Mexico City, 1903), 11–13.

25. Ramón Sánchez Flores, *Historia de la tecnología y la invención: introducción a su estudio y documentos para los anales de la técnica* (Mexico City: Fomento Cultural Banamex, 1980), 291.

26. Crespo Martínez, *Las patentes de invención*, 39.

27. Elías Trabulse, *Las patentes de invención durante el siglo XIX en México*, Boletín del Archivo General de la Nación, 3rd series, no. 34 (Mexico City: Boletín del Archivo General de la Nación, 1988).

28. Enrique Hernández Aranda, un perfeccionamiento en la fabricación de cerillos, bujías de seguridad, February 16, 1885, AGN: Patentes y Marcas, caja 28, exp. 1263, fs. 1–34.

29. For a farmer to increase his profit margins, he had to plant new crops or experiment with new agricultural techniques and equipment, and these seemingly minor experiments came at a high risk, perhaps even the risk of starvation. Beezley, *Judas at the Jockey Club*, 67–88.

30. "Necesito capital para invento," *El Diario*, December 29, 1900, 8.

31. Edmund Downey, "Behind the Door," *Mexican Herald*, November 22, 1895, 3–4.

32. Tenorio-Trillo, *Mexico at the World's Fairs*, 134.

33. Edward N. Beatty and Lucrecia Orensanz Escofet, "Invención e innovación: ley de patentes y technología en el México del siglo XIX," *Historia Mexicana* 45, no. 3 (1996): 599.

34. Howard Conkling, *Mexico and the Mexicans: Or, Notes of Travel in the Winter and Spring of 1883* (New York: Taintor Brothers, Merrill and Co., 1883), 232.

35. *Patent Law of the United States of Mexico: Law for the Promotion of New Industries in the United States of Mexico; Trade Mark Law of the United States of Mexico.* For similar pamphlets see Antonio Enciso Ulloa, H. G. Perkins, and Louis C. Simonds, *The Mexican Pocket Lawyer* (Mexico City: Imprenta de Hull, 1910), 66–102.

36. Clemens Graaff, un aparato para apagar incendios, April 12, 1911, AGN: Patentes y Marcas, leg. 308, exp. 3, fs. 1–36; Domingo Biosea Galcerán, perfeccionamiento de aparato extintor de incendios, November 19, 1904, AGN: Patentes y Marcas, leg. 308, exp. 10, fs. 1–3; Frank Vanoy Carman, mejoras para hidrantes para incendios, September 19, 1907, AGN: Patentes y Marcas, leg. 308, exp. 17, fs. 1–6.

37. Crespo Martínez, *Las patentes de invención*, 67.

38. Mark Jefferson, "The Geographic Distribution of Inventiveness," *Geographical Review* 19, no. 4 (1929): 653. He also conducted a study of sixty-nine nations' cultural progress, measured by the number of schools, railroads, exports, and mail system. Mexico ranked number thirty-nine in his study of culture; see "The Culture of the Nations," *Bulletin of the American Geographical Society* 43, no. 4 (1911): 256.

39. Crespo Martínez, *Las patentes de invención*, 67–69.

40. Jefferson, "Geographic Distribution of Inventiveness," 649.

41. Mark D. Janis, "Patent Abolitionism," *Berkeley Technology Law Journal* 17, no. 2 (2002): 916; Christine MacLeod, "The Invention of Heroes," *Nature* 460, no. 30 (2009): 572–73; Thomas P. Hughes, *American Genesis: A Century of Invention and Technological Enthusiasm, 1870–1970* (Chicago: University of Chicago Press, 1989), xv–xvi.

42. During this age of heroic inventors, popular literature and biographies praised amateur inventors and highlighted the exceptional rags-to-riches stories; see Thomas P. Hughes, *Elmer Sperry: Inventor and Engineer* (Baltimore, MD: Johns Hopkins University Press, 1971); Christine MacLeod, "Concepts of Invention and the Patent Controversy in Victorian Britain," in *Technological Change: Methods and Themes in the History of Technology*, ed. Robert Fox (Amsterdam: Harwood Academic, 1996), 137–38.

43. Hughes, *American Genesis*, xv–xvi.

44. Díaz y Sala, una composición para extinguir fuego de cualquier naturaleza, April 18, 1905, AGN: Patentes y Marcas, leg. 308, exp. 12, fs. 1; Rubén Martí, un extinguidor para incendios, January 21, 1908, AGN: Patentes y Marcas, leg. 308, exp. 19, fs. 1–3; Reinaldo Rodríguez Arce and Everardo Rodríguez Arce, un procedimiento para extinguir incendios en pozos de petróleo o gas, July 29, 1908, AGN: Patentes y Marcas, leg. 308, exp. 20, fs. 1–4.

45. Díaz y Sala, una composición de materias para extinguir las flamas en los incendios denominada "mata flama," July 8, 1910, AGN: Patentes y Marcas, leg. 308, exp. 27, fs. 1.

46. M. Cisneros, un procedimiento para la fabricación de un techo impermeable, April 15, 1905, AGN: Patentes y Marcas, leg. 285, exp. 47.

47. Carmen Chávez, Fabricación de una pasta impermeable denominada invulnerable, AGN: Patentes y Marcas, leg. 285, exp. 58.

48. Gamboa, *Santa*, 116–17.

49. Frederick Smyth, *Discourses and Letters Commemorative of Emily Lane, Wife of Ex-Gov. Frederick Smyth* (Manchester, NH: John B. Clarke, 1885), 108.

50. Daniel Blumenkron, cerillos impermeables, October 28, 1878, AGN: Patentes y Marcas, caja 14, exp. 848, fs. 2; Daniel Blumenkron, cerillos impermeables, September 26, 1879, AGN: Patentes y Marcas, caja 14, exp. 848, fs. 1–8.

51. Daniel Blumenkron, Manufacture of Matches, U.S. Patent no. 242,427, issued June 7, 1881.

52. *Anales del Ministerio de Fomento de la República Mexicana, Tomo IV* (México: Imprenta de Francisco Díaz de Leon, 1881), 120–123; Mariano Barcena, *La 2. Exposición de 'las clases productoras,' y descripción de la ciudad de Guadalajara* (Guadalajara: Tipografía de Sinforoso Banda, 1880), 177 and 180; Miguel Ulloa, *Memoria de la primera exposición en la capital del estado de México, Toluca* (México: Tipografía Literaria del Filomeno Mata, 1883), 85 and 188.

53. "Incendio," *Patria*, July 30, 1880, 2.

54. "Ayuntamiento el de Puebla ha quedado formado," *El Siglo Diez y Nueve*, September 13, 1880, 3; "Cámara de comercio de Puebla," *El Siglo Diez y Nueve*, March 16, 1893, 2.

55. "Death of Dr. P. Marin," *Mexican Herald*, March 31, 1896, 2.

56. Beginning in 1900, General Electric, American Telegraph & Telephone, Du Pont, and Kodak began to employ scientists in their research and development teams in order to protect their companies against future competition. David A. Hounshell and John Kenly Smith, Jr., *Science and Corporate Strategy: Du Pont R&D, 1902–1980* (New York: Cambridge University Press, 1988).

57. Beatty, "Approaches to Technology Transfer," 192.

58. Beatty, "Patents and Technological Change," 138.

59. Merritt Roe Smith and Leo Marx, *Does Technology Drive History? The Dilemma of Technological Determinism* (Cambridge, MA: MIT Press, 1994), 38. Donald MacKenzie's study of nuclear missiles shows that institutions and politics, stemming from a specific social context, determined what technological improvements needed to be made; see *Inventing Accuracy: A Historical Sociology of Nuclear Missile Guidance* (Cambridge, MA: MIT Press, 1990).

60. Enrique Hernández Aranda, Un perfeccionamiento en la fabricación de cerillos, bujías de seguridad, 1885–1886, AGN: Patentes y Marcas, caja 28, exp. 1263, fs. 1.

61. Porfirio Parra, "The General Character of the Positive Method," in *Nuevo sistema de lógica inductiva y deductiva* (Mexico City: Tipografía Ecónomica, 1903), 225.

62. Oliver C. Ellis, *A History of Fire and Flame 1932* (London: J. W. Northend, 1932), 69.

63. Enrique Hernández Aranda, Un perfeccionamiento en la fabricación de cerillos, bujías de seguridad, 1885–86, AGN: Patentes y Marcas, caja 28, exp. 1263, fs. 17.

64. Francisco Tadeo Linder, fabricación de cerillos, fósforos y yesca de seguridad, November 27, 1858, AGN: Patentes y Marcas, caja 5, exp. 369.

65. Adolfo Martínez Urista, un extinguidor de incendios, December 28, 1910, AGN: Patentes y Marcas, leg. 308, exp. 33, fs. 1. Martínez Urista also patented an invention to prevent boilers from getting clogged with debris: Composition for Preventing Scale in Boilers, U.S. Patent no. 950,582, application filed May 21, 1909.

66. Claudia Agostoni, *Monuments of Progress: Modernization and Public Health in Mexico City, 1876–1910* (Boulder: University Press of Colorado, 2003), xii and 23; Wakild, "Naturalizing Modernity," 110.

67. Marketing ideas of health and healthiness became a common motif in advertisements, and Sunkist orange campaigns offer a clear example of this strategy; see Douglas Cazaux Sackman, *Orange Empire: California and the Fruits of Eden* (Berkeley: University of California Press, 2005), 109.

68. Daniel Blumenkron, Cerillos impermeables, October 28, 1878, AGN: Patentes y Marcas, caja 14, exp. 848, fs. 2.

69. E. M. Arzac to the Ciudadano Secretario de Fomento Colonización Industria y Comercio, February 1, 1886, AGN: Patentes y Marcas, fs. 23.

70. Melosi, *Sanitary City*, 423.

71. Agustín Rousseau y Luis Chaubet, Fabricación de fósforos y cerillos inofensivos, 1858–1859, AGN: Patentes y Marcas, caja 4, exp. 366, fs. 1; Claudia Clark, *Radium Girls: Women and Industrial Health Reform, 1910–1935* (Chapel Hill: University of North Carolina Press, 1997); Melvin L. Myers and James D. McGlothlin, "Matchmakers' 'Phossy Jaw' Eradicated," *American Industrial Hygiene Association Journal* 57, no. 4 (1996): 330–32.

72. Barbara Sicherman, *Alice Hamilton: A Life in Letters* (Cambridge, MA: Harvard University Press, 1984), 155–56.

73. Luis Mendez, *Gaceta de los Tribunales de la República Mexicana*, vol. 3 (México: Isidoro Devaux, 1862), 633.

74. Sidney Chalhoub, "The Politics of Disease Control: Yellow Fever and Race in Nineteenth-Century Rio de Janeiro," *Journal of Latin American Studies* 25, no. 3 (1993): 441–43; Melosi, *Sanitary City*, 17–42; Libby Hill, *The Chicago River: A Natural and Unnatural History* (Chicago: Lake Claremont Press, 2000), 112; Julie Sze, *Noxious New York: The Racial Politics of Urban Health and Environmental Justice* (Cambridge, MA: MIT Press, 2007), 32–33; Petri S. Juuti, Tapio S. Katko, and Heikki S. Vuorinen, *Environmental History of Water: Global Views on Community Water Supply and Sanitation* (London: IWA Publishing, 2007), 106–10.

75. Discussions about installing a telephone in the fire department came about as early as 1882, "Boletin del 'Monitor,'" *El Monitor Republicano*, September 28, 1882, 1; Víctor Cuchí Espada, "Cambio de costumbres o cómo ser moderno: comerciantes, Ayuntamiento y mercado telefónico en la Ciudad de México, 1881–1905," *Cuicuilco* 15 (1999): 265–303.

76. Matilde Rábago Vda. de Ovieda, Un aparato anunciador de incendios, February 16, 1911, AGN: Patentes y Marcas, leg. 308, exp. 28, fs. 1–2.

77. Miguel Laimón, Un auto anunciador de incendio, April 23, 1907, AGN: Patentes y Marcas, leg. 308, exp. 16, fs. 1–2.

78. Guillermo Taverner, Alarma y auto-extinguidor de incendios "Taverner," September 1, 1909, AGN: Patentes y Marcas, leg. 308, exp. 26, fs. 1–7.

79. H. F. Neefus, *Sprinkler Equipments: Installation and Requirements* (Newark, NJ: The Merchants' Insurance Co., 1895), 8–11.

80. American School of Correspondence, *Cyclopedia of Fire Prevention and Insurance*, 88.

81. "Organization of the Police Department in Mexico," *Overland Monthly: The Awakening of Mexico Centenary of the Republic* 56 (July 1910): 117.

82. Sobre que se reconozca si el betún o asfalto que cubre la parte exterior del techo del teatro de la 4a Calle de Relox, propriedad de D. José Soledad Aycardo, es peligrosa a un incendio, April 2, 1866, AHDF: AGDF, Teatros, vol. 4017, exp. 73.

83. Carlos Villegas, December 6, 2007, AGN: Patentes y Marcas, leg. 308, exp. 18, fs. 1–2.

84. Ricardo Rojas, Un aparato automático contra incendio de películas cinematográficas, June 29, 1912, AGN: Patentes y Marcas, leg. 308, exp. 35, fs. 1–3.

85. Reinaldo Rodríguez Arce and Everardo Rodríguez Arce, Un procedimiento para extinguir incendios en pozos de petróleo o gas, July 29, 1908, AGN: Patentes y Marcas, leg. 308, exp. 20, fs. 1–4.

86. Ernesto Fuchs, Un procedimiento para extinguir incendios de pozos de petróleo, August 11, 1908, leg. 308, exp. 21, fs. 1; J. Figueroa Doménech, *Guía general descriptive de la República Mexicana*, vol. 2 (Barcelona: R. de S.N. Araluce, 1899), 253; *Journal of the American Society of Mechanical Engineers* 30, no. 7 (1908): 23.

87. Nathan Rosenberg, "Historiography of Technical Progress," 19.

88. Incendio ocurrido en la panadería del Pte. Quebrado no. 7, April 20, 1907, AHDF: AGDF, Fábricas, vol. 1604, exp. 356, fs. 1.

89. Concejal Ingeniero Abraham Chávez to C. Secretario del Consejo S. del Gobierno del Distrito, February 7, 1912, AHDF: AGDF, Diversiones Públicas, vol. 807, exp. 1323; Inspector General de Policía to Jefe del Cuerpo de Bomberos, August 2, 1912, AHDF: AGDF, vol. 1396, exp. 1053; Concejal Ingeniero Abraham Chávez to C. Secretario del Consejo S. del Gobierno del Distrito, October 30, 1912, AHDF: AGDF, Diversiones Públicas, vol. 596, exp. 21.

90. Inspector General to C. Gobernador del Distrito, July 6, 1916, AHDF: AGDF, Gobierno del Distrito, Mercados, vol. 1745, exp. 1288, fs. 1.

91. Paul A. David, *Technological Choice Innovation and Economic Growth: Essays on American and British Experience in the Nineteenth Century* (Cambridge: Cambridge University Press, 1975); Jacob Schmookler, *Invention and Economic Growth* (Cambridge, MA: Harvard University Press, 1966).

92. Edward N. Beatty and Lucrecia Orensanz Escofet make a distinction between *invention* and *innovation*. *Invention* is the creation of something entirely new, whereas *innovation* is the adaptation and improvement of existing technologies. See "Invención e innovación," 570.

93. Abbott Payson Usher, *A History of Mechanical Inventions*, 2nd ed. (Cambridge, MA: Harvard University Press, 1954); Rosenberg, "Historiography of Technical Progress," 19.

94. "Solicitud de Privilegio," *El Monitor Republicano*, August 27, 1884, 3; "Jarabe depurativo antisifilitico de Juan Septien," *El Monitor Republican*, August 20, 1882, 4.

95. Juan Septien, Un medio para extinguir incendios, August 20, 1884, AGN: Patentes y Marcas, caja 25, exp. 1147, fs. 1–9.

96. Wadsworth, Martínez, Longman, Unas granadas de mano para extinguir incendios, September 2, 1884, AGN: Patentes y Marcas, caja 25, fs. 8–19.

97. Wadsworth, Martínez, Longman, Unas granadas de mano para extinguir incendios, September 2, 1884, AGN: Patentes y Marcas, caja 25, fs. 9–10.

98. Wadsworth, Martínez, and Longman to Don Pedro Garay, July 3, 1884, AGN: Patentes y Marcas, caja 25, fs. 13.

99. Charles John Hexamer, *Fire Hazards in Textile Mills, Mill Architecture, and Means for Extinguishing Fire: Three Lectures Delivered before the Franklin Institute* (Philadelphia: Merrihew Print, 1885), 58.

100. *The Harden Hand Grenade Fire Extinguisher, Patd Aug. 8, 1871. Patd. Aug. 14, 1883* (Chicago: The Harden Hand Grenade Fire Extinguisher Company, 1884), Hagley Library Trade Catalogues.

101. "La major protección absolutamente contra el fuego," *El Siglo Diez y Nueve*, December 26, 1874, 4.

102. Daniel J. Kenny, *Cincinnati Exposition Guide and Catalogue of the Fine Arts Department, Containing the Name and Address of Every Exhibitor at the Fifth Cincinnati Industrial Exposition of 1874* (Cincinnati, OH: Cincinnati Gazette Co., 1874).

103. Babcock Manufacturing Company, *The Babcock Self-Acting Portable Copper Fire Extinguisher* (Chicago: Babcock Manufacturing Company, 1875); S. F. Hayward, *The Improved Babcock Fire Extinguisher* (Boston, New England Office, 1882), Hagley Library Trade Catalogues.

104. Minimax brand fire-extinguishers by la casa Brandestein y Compañía, March 4, 1908, AGMP, tomo 480, ficha 16526, leg. 23, exp. 23, letra B, fs. 49–82.

105. W. Graaf & Compagnie, Gesellschaft mit besehränkter Haftung, ciertas mejoras en extinguidores de incendio, February 21, 1905, AGN: Patentes y Marcas, leg. 308, exp. 11, fs. 1–4; W. Graaf & Compagnie, G.m,b,H, extinguidores de incendio perfeccionados, October 16, 1906, AGN: Patentes y Marcas, leg. 308, exp. 13, fs. 1–5.

106. Joseph J. Corn, *User Unfriendly: Consumer Struggles with Personal Technologies, from Clocks and Sewing Machines to Cars and Computers* (Baltimore, MD: Johns Hopkins University Press, 2011).

107. Rubén Martí, Un extinguidor para incendios, January 21, 1908, AGN: Patentes y Marcas, leg. 308, exp. 19, fs. 1–3.

108. S. F. Hayward & Company, "Rescue Fire Extinguisher," 1916, Hagley Library Trade Catalogues, 0473, file 1058.

109. "Chemical Soda and Acid Fire Extinguisher," *Municipal Supplies: Fire Protection, Waterworks, Street-Police-Sewers, catalog no. 50* (Chicago: W. S. Darley and Co., 1920), Hagley Library Trade Catalogues, 2280, box 12, fldr. 3, p. 24.

110. "S. F. Hayward and Co., to Joseph Durland," Chester Historical Society: W. S. Durland Collection, October 1890.

111. Domingo Biosea Galcerán, Perfeccionamiento de un aparato extintor de incendios, November 19, 1904, AGN: Patentes y Marcas, leg. 308, exp. 1, fs. 1–3; Díaz y Sala, Una composición de materias para extinguir las flamas en los incendios denominada "mata flama," July 8, 1910, AGN: Patentes y Marcas, leg. 308, exp. 27, fs. 1.

112. William H. Beezley, "Mexican Sartre on the Zócalo: Nicolás Zúñiga y Miranda," *The Human Tradition in Latin America: The Nineteenth Century*, ed. Judith Ewell and William H. Beezley (Wilmington, DE: Scholarly Resources, 1989), 204–14.

113. "Fiend Holmes Must Go to Trial. Charged with Numerous Cold Blooded Murders Committed to Obtain Life Insurance," *Mexican Herald*, October 28, 1895, 1.

114. Jeffrey Pilcher, *The Sausage Rebellion: Public Health, Private Enterprise, and Meat in Mexico City, 1890–1917* (Albuquerque: University of New Mexico Press, 2006), 75–76; Jeffrey Pilcher, "Mad Cowmen, Foreign Investors and the Mexican Revolution," *Journal of Iberian and Latin American Studies* 4, no. 1 (1998): 3.

115. Beatty, "Approaches to Technology Transfer," 180.

116. Beezley, "Mexican Sartre on the Zócalo," 210–11.

117. "City of Mexico," *Sanborn Fire Insurance Maps*, 1905, Perry-Castañeda Library Map Collection, University of Texas Libraries.

CHAPTER SIX. INSURING PROGRESS

Epigraph: National Board of Fire Underwriters, *Fire Insurance: Its Importance; Its Relation to the Community* (New York: Press of Styles and Cash, 1899), 28.

1. A. F. Dean, "Fire Insurance," in *Lectures on Commerce Delivered before the College of Commerce and Administration of the University of Chicago*, ed. Henry Rand Hatfield (Chicago: University of Chicago Press, 1907), 322.

2. J. A. Fowler, *The Fire Account—Physical, Personal, Moral. Address Delivered before the Fire Underwriters' Association of the Northwest* (Chicago, 1878), 5.

3. Quoted in National Board of Fire Underwriters, *Fire Insurance*, 28.

4. National Board of Fire Underwriters, *Fire Insurance*, 27–28.

5. Dalit Baranoff, "Shaped by Risk: The American Fire Insurance Industry, 1790–1920" (PhD diss., Johns Hopkins University, 2004), 4–6.

6. Reynaldo Sordo Cedeño, "La sociedades de socorros mutuos, 1867–1880," *Historia Mexicana* 33, no. 1 (1983): 72–96.

7. Juan Félipe Leal, *Del mutualismo al sindicalismo: 1843–1911* (Mexico City: Ediciones El Caballito, 1991), 13–14.

8. Georgina H. Endfield explains how vulnerability is anything but static. Social, economic, and political factors determine what makes societies more or less vulnerable; see *Climate and Society in Colonial Mexico: A Study of Vulnerability* (Malden, MA: Blackwell Publishing, 2008), 4–5; Javier Auyero and Débora Alejandra Swistun, *Flammable: Environmental Suffering in an Argentine Shantytown* (Oxford: Oxford University Press, 2009); Susan Cutter, "Vulnerability to Environmental Hazards," *Progress in Human Geography* 20, no. 4 (1996): 529–39.

9. *La Fraternal Compañía de Seguros de Vida y Accidentes* 4, no. 46 (October 31, 1897), 1, UNAM, Fondo Reservado [hereafter FR], microfilm.

10. Dean, "Fire Insurance," 326–27.

11. Fowler, *Fire Account*, 6.

12. W. H. Frazier, *Few and Furious: Also, a Reprint of the Second Edition of the Little Paragraphs: Comments and Criticisms on the Business of Fire Insurance* (Philadelphia, 1902), 22.

13. Royal Insurance Company, *Index to Classification of Fire Hazards: North America*. S. L.: S. N., 1897, Hagley Library.

14. Frazier, *Few and Furious*, 5.

15. F. C. Moore, *Fires: Their Causes, Prevention and Extinction, Combining Also a Guide to Agents Respecting Insurance against Loss by Fire* (New York: Continental Insurance Co., 1877), 201 and 209.

16. Dean, "Fire Insurance," 359.

17. American School of Correspondence, *Cyclopedia of Fire Prevention and Insurance*, 92.

18. Bird, *Protection against Fire*, 13.

19. Henry Daniel of Bristol quoted in J. A. Fowler, *Fire Account*, 14.

20. Arthur C. Ducat, *The Practice of Fire Underwriting*, 4th ed. (New York: T. Johns Jr., 1866), 12.

21. Barry Supple, "Insurance in British History," in *The Historian and the Business of Insurance*, ed. Oliver M. Westall (Manchester: Manchester University Press, 1984), 2–3.

22. *La Fraternal Compañía de Seguros de Vida y Accidentes* 1, no. 9 (September 20, 1894), UNAM, FR, microfilm.

23. D. T. Jenkins, "The Practice and Insurance against Fire, 1750–1840, and Historical Research," in *The Historian and the Business of Insurance*, ed. Oliver M. Westall (Manchester: Manchester University Press, 1984), 18–19.

24. *La Previsora y La Bienhechora compañías de seguros mutuos contra incendio y sobre la vida* (Mexico City: Imprenta de Andrade y Escalante, 1865), 4.

25. *La Previsora y La Bienhechora compañías de seguros mutuos contra incendio y sobre la vida* (Mexico City: Imprenta de Andrade y Escalante, 1865).

26. *La Previsora y La Bienhechora compañías de seguros mutuos contra incendio y sobre la vida* (Mexico City: Imprenta de Andrade y Escalante, 1865), 5.

27. *Boletín de las leyes del Imperio Mexicano ó sea codigo de la restauración*, vol. 4 (Mexico City: José Sebastian Segura, 1865), 303.

28. *Memoria presentada á S. M. el Emperador por el Ministro de Fomento Luis Robles Pezuela de los trabajos ejecutados en su ramo el año de 1865* (Mexico City: Imprenta de J. M. Andrade y F. Escalante, 1866), 83–85.

29. Antonio Minzoni Consorti, *Crónica de dos siglos de seguro en México* (Mexico City: Comisión Nacional de Seguros y Fianzas, 2005), 22–23.

30. *The Massey-Gilbert Blue Book of Mexico: A Directory in English of Mexico City* (International Bank and Trust Company of America, 1903), 349–51.

31. *La Fraternal Compañía de Seguros de Vida y Accidentes* 1, no. 7 (July 21, 1894) UNAM, FR, microfilm, fs. 4.

32. Minzoni Consorti, *Crónica de dos siglos de seguro en México*, 24; *La Fraternal Compañía de Seguros de Vida y Accidentes* 1, no. 8 (August 31, 1894), UNAM, FR, microfilm, fs. 3.

33. Minzoni Consorti, *Crónica de dos siglos de seguro en México*, 31.

34. Luís A. Vidal y Flor, *Colección de Leyes Federales Vigentes sobre Instituciones de Crédito, Ferrocarriles, Compañías de Seguros, Almacenes generals de Depósito y varias Circulares importantes recientes* (Mexico City: Herrero Hermanos, 1900), 165.

35. Minzoni Consorti, *Crónica de dos siglos de seguro en México*, 44.

36. *La Fraternal Compañía de Seguros de Vida y Accidentes* 1, no. 9 (September 30, 1894), UNAM, FR, microfilm, fs. 1.

37. Minzoni Consorti, *Crónica de dos siglos de seguro en México*, 22.

38. *La Fraternal Compañía de Seguros de Vida y Accidentes* 1, no. 1 (January 31, 1894), UNAM, FR, microfilm, fs. 1.

39. *La Fraternal Compañía de Seguros de Vida y Accidentes* 2, no. 13 (January 31, 1895), UNAM, FR, microfilm, fs. 1.

40. *La Fraternal Compañía de Seguros de Vida y Accidentes* 1, no. 1 (January 31, 1894), UNAM, FR, microfilm, fs. 1.

41. *La Fraternal Compañía de Seguros de Vida y Accidentes* 1, no. 1 (January 31, 1894), UNAM, FR, microfilm, fs. 1.

42. "Viudas y Huérfanos," *La Fraternal Compañía de Seguros de Vida y Accidentes* 4, no. 46 (October 31, 1897), UNAM, FR, microfilm, fs. 1.

43. "Viudas y Huérfanos," *La Fraternal Compañía de Seguros de Vida y Accidentes* 4, no. 46 (October 31, 1897), UNAM, FR, microfilm, fs. 1; "La Fraternal: Compañía de Seguros de Vida y Accidentes," *El Siglo Diez y Nueve*, April 7, 1894, 4; "Seguro Contra Incendio. La Compañía Hamburguesa-Bremesa de Seguros Contra Incendio," *El Siglo Diez y Nueve*, November 20, 1867, 4; "Compañía Alemana de Seguros Contra Incendio. Titulada: Deutsche Feuer Versicherungs-Actién-Gesellschaft zu Berlin," *Siglo Diez y Nueve*, September 26, 1871, 4.

44. "Terrible Incendio," *La Fraternal Compañía de Seguros de Vida y Accidentes* 8, no. 87 (March 31, 1901), UNAM, FR, microfilm, fs. 2.

45. *La Fraternal Compañía de Seguros de Vida y Accidentes* 2, no. 13 (January 21, 1895), UNAM, FR, microfilm, fs. 1.

46. *La Fraternal Compañía de Seguros de Vida y Accidentes* 1, no. 10 (October 1894), UNAM, FR, microfilm, fs. 1.

47. "Compañías de Seguros Contra Incendio Central Insurance, Home Insurance, British Crown, Phoenix," December 23, 1911, AGN: Tribunal Superior de Justicia del Distrito Federal [hereafter TSJDF], caja 1056, fs. 196.

48. *La Fraternal Compañía de Seguros de Vida y Accidentes* 1, no. 7 (July 21, 1894), UNAM, FR, microfilm, fs. 4.

49. "Compañías de Seguros Contra Incendio Central Insurance, Home Insurance, British Crown, Phoenix," December 23, 1911, AGN: TSJDF, caja 1056, fs. 196.

50. "Incendio de la fábrica de dulces 'Los Pirineos,'" *El Nacional*, October 23, 1895, 2.

51. "Amparo. Termino para interponer el recurso alegaciones presentadas a la Suprema Corte de Justicia bajo el patrocinio del Licenciado José Diego Fernandez. Por las compañías de seguro en el amparo que tiene pedido con-

tra una sentencia de la 4a sala de tribunal superior," March 5, 1898, AGN: Archivo Histórico de la Suprema Corte de Justicia de la Nación, Tribunal Pleno, exp. 451, fs. 3–4.

52. "Sentencia Pronunciada por la 4a Sala del Tribunal Superior del Distrito en el Juicio Segundo por el Señor Don Juan Maria Colomic con las Compañías de Seguros Contra Incendio 'North British and Mercantile' y 'Liverpool and London and Globe,' sobre pago de un siniestro." July 28, 1897, AGN: Archivo Histórico de la Suprema Corte de Justicia de la Nación, Tribunal Pleno, exp. 451, fs. 8–13.

53. "Sentencia Pronunciada por la 4a Sala del Tribunal Superior del Distrito en el Juicio Segundo por el Señor Don Juan Maria Colomic con las Compañías de Seguros Contra Incendio 'North British and Mercantile' y 'Liverpool and London and Globe,' sobre pago de un siniestro." July 28, 1897, AGN: Archivo Histórico de la Suprema Corte de Justicia de la Nación, Tribunal Pleno, exp. 451, fs. 9.

54. "Sr. Pablo Alexanderson y Emilio Eberle a La Secretaria de la Suprema Corte de Justicia de la Nación," May 11, 1898, AGN: Archivo Histórico de la Suprema Corte de Justicia de la Nación, Tribunal Pleno, exp. 451, fs. 1.

55. "Silverio Fernández en representación de la Sociedad Fornaguera Ortíz y Cia," December 12, 1903, AGN: TSJDF, caja 0293, fs. 116.

56. Los Heróicos Boeros, March 24, 1902, AGN: TSJDF, caja 0154, fs 1–10.

57. Pasta de la Magdeburguesa Seguros contra indendio. Chrislieb & Rubke de México, 1907, AGN: Comisión Monetaria, caja 870.

58. Uwe Lübken and Christof Mauch, "Uncertain Environments: Natural Hazards, Risk, and Insurance in Historical Perspective," *Environment and History* 17, no. 1 (2011): 9.

59. Jason A. Gilliland and Mathew Novak, "On Positioning the Past with the Present: The Use of Fire Insurance Plans and GIS for Urban Environmental History," *Environmental History* 11, no. 1 (2006): 136–39; Thomas Krafft, "Reconstructing the North American Urban Landscape: Fire Insurance Maps—An Indispensable Source (Feuerversicherungskarten: Unverzichtbares Hilfsmittel zur historisch—geographischen Rekonstruktion der nordamerikanischen Stadt)," *Erdukunde* 47, no. 3 (1993): 196–211.

60. R. J. Hayward, "Insurance Plans and Land Use Atlases: Sources for Urban Historical Research," *Urban History Review* 2 (1973): 2–9.

61. "Key Plan of City of Mexico, Mexico," *Sanborn Fire Insurance Maps*, 1905, Perry-Castañeda Library Map Collection, University of Texas Libraries.

62. Barbara J. Williams, "Tepetate in the Valley of Mexico," *Annals of the Association of American Geographers* 62, no. 4 (1972): 618–26.

63. "City of Mexico, Plan 11," *Sanborn Fire Insurance Maps*.

64. "City of Mexico, Plan 3," *Sanborn Fire Insurance Maps.*

65. "City of Mexico, Plan 7," *Sanborn Fire Insurance Maps.*

66. Thomas Philip Terry, *Terry's Mexico: Handbook for Travellers* (London: Gay and Hancock, 1911), 231.

67. "City of Mexico, Plan 13," *Sanborn Fire Insurance Maps.*

68. "City of Mexico, Plan 2," *Sanborn Fire Insurance Maps.*

69. Alfred Oscar Coffin, *Land without Chimneys; or, The Byways of Mexico* (Cincinnati: Editor Publishing, 1898), 14.

70. American School of Correspondence, *Cyclopedia of Fire Prevention and Insurance*, 92.

CHAPTER SEVEN. HEALING THE HAZARDOUS CITY

1. The places where people live shape their overall wellness, making environments and health inseparable. See Rachel Carson, *Silent Spring* (Boston: Houghton Mifflin Company, 1962), 1–6; Linda Nash, *Inescapable Ecologies: A History of Environment, Disease, and Knowledge* (Berkeley: University of California Press, 2006), 1–5; Conevery Bolton Valenčius, *The Health of the Country: How American Settlers Understood Themselves and Their Land* (New York: Basic Books, 2002), 4–10.

2. One of the earliest historians of Latin America to link health and state formation was Nancy Leys Stephan in *The Hour of Eugenics: Race, Gender, and Nation in Latin America* (Ithaca, NY: Cornell University Press, 1991). Several recent studies have continued to examine the connection between health and national identity. See José G. Amador, "'Redeeming the Tropics': Public Health and National Identity in Cuba, Puerto Rico, and Brazil, 1890–1940" (PhD diss., University of Michigan, 2008); Ivette Rodriguez-Santana, "Conquests of Death: Disease, Health and Hygiene in the Formation of a Social Body (Puerto Rico, 1880–1929)" (PhD diss., Yale University, 2005).

3. Aréchiga Córdoba, "La ciudad y Tepito," 171–207; Diego Armus, *The Ailing City: Health, Tuberculosis, and Culture in Buenos Aires, 1870–1950* (Durham, NC: Duke University Press, 2011); Patricio V. Márquez and Daniel J. Joly, "A Historical Overview of the Ministries of Public Health and Medical Programs of the Social Security Systems in Latin America," *Journal of Public Health Policy* 7, no. 3 (1986): 378–94.

4. Anne-Emanuelle Birn and Raúl Necochea López recently noted that occupational health in Latin America has been overlooked; see "Footprints on the Future: Looking Forward to the History of Health and Medicine in Latin America in the Twenty-First Century," *HAHR* 91, no. 3 (2011): 517. Exceptions include Angela Vergara, "The Recognition of Silicosis: Labor Unions and Physicians in the Chilean Copper Industry, 1930s–1960s," *Bulletin of the History of*

Medicine 79, no. 4 (2005): 723–48; Anna Beatriz de Sá Almeida, "A Associação Brasileira de Medicina do Trabalho: Locus do processo de constituição da especialidade medicina do trabalho no Brasil na década de 1940," *Ciência e Saúde Coletiva* 13, no. 3 (2008): 869–77.

5. Brett L. Walker, *Toxic Archipelago: A History of Industrial Disease in Japan* (Seattle: University of Washington Press, 2010), 6.

6. Natalia Priego, *Science, Culture and Society in Mexico 1860–1940: The Contradictions of the Quest for Modernity* (Saarbrücken, Germany: VDM Verlag Dr. Müller Aktiengesellschaft and Co. KG, 2009), 50.

7. Martha Eugenia Rodríguez, *La escuela nacional de medicina, 1833–1910* (Mexico City: Universidad Nacional Autónoma de México, 2008), 9.

8. Federico García Sepúlveda, "Estadística general del Hospital Juárez, 1888–1895" (MD thesis, Escuela de Medicina, 1896), 33–34. For some international perspective, in 1858 in the United Kingdom, 3,125 people died from burns. See "Deaths from the Inflammability of Clothing," *Lancet* 76, no. 1932 (1860): 245. Mexico's Superior Health Council also published an annual account of mortality rates in the capital and recorded that between the years 1879–1882, thirty-five people died of burn-related causes; see *Boletín del Consejo Superior de Salubridad del Distrito Federal. Julio de 1880 a Junio de 1881*, vol.1 (Mexico City: Imprenta del Gobierno, En Palacio. Dirigida por Sabás A. y Munguía, 1881).

9. Cátedra de Materia Médica, Farmacología y Terapeutica. Programa del curso para el año de 1895, Archivo Histórico de Facultdad de Medicina [hereafter AFM]: Fondo Escuela de Medicina y Alumnos [hereafter FEMYA], leg. 192, exp. 1, fs, 78; Cátedra de Medicina Operatoria. Programa para el año de 1903, AFM: FEMYA, leg. 192, exp. 2, fs. 8–9.

10. "Programa: propuesto por el profesor de 2° año de Clinica Externa para dar el curso durante el año escolar de 1895," December 24, 1894, AFM: FEMYA, leg. 192, exp. 1, fs. 82.

11. Rodríguez, *La escuela nacional de medicina*, 83–84.

12. Auguste Nélaton, *Élémens de pathologie chirurgicale* (Paris: Germer Baillière, Librairie-Éditeur, 1844), 286–97.

13. Nélaton, *Élémens de pathologie chirurgicale*, 290–95.

14. Castillo Velasco, *Colección de leyes*, 221.

15. Arpad G. Gerster, *The Rules of Aseptic and Antiseptic Surgery: A Practical Treatise for the Use of Students and the General Practitioner* (New York: D. Appleton and Company, 1890), vii.

16. Gerster, *Rules of Aseptic and Antiseptic Surgery*, 11–12; Henri de Rothschild, *Tratamiento de las Quemaduras por el método céreo (cura por la ambrina)*, trans. D. José de Sard (Barcelona: Casa Editorial P. Salvat, 1919), 38.

17. Pedro Martínez Garza, "Apuntes de clínica externa" (MD thesis, Escuela de Medicina, 1895), 45.

18. Thomas Baynton, *Descriptive Account of a New Method of Treating Old Ulcers of the Legs*, 2nd ed. (London: Emery and Adams, 1799), 21–23.

19. Edward Greenhow, "New Method of Treating Burns," *London Medical Gazette* 1 (1839): 82–84.

20. John Fisher and Natalie Priego, "Ignorance and 'Habitus': Blinkered and Enlightened Approaches towards the History of Science in Latin America," *Bulletin of Latin American Research* 25, no. 4 (2006): 533.

21. Manuel Andrade, "Nuevo tratamiento de la quemaduras," *La Gaceta Médica de México* 3 (1838): 347–52.

22. Rothschild, *Tratamiento de las quemaduras*, 42.

23. "Interesante descubrimiento," *El Siglo Diez y Nueve*, May 21, 1896, 2.

24. M. Felix Freshwater and Thomas J. Krizek, "George David Pollock and the Development of Skin Grafting," *Annals of Plastic Surgery* 1, no. 1 (1978): 96–104; Jacques-Louis Reverdin, "Greffe épidermique," *Bulletin de la Société Impériale de Chirurgie Paris* 10 (1869): 511–15.

25. "Alrededor del mundo: ingertos de piel," *El Imparcial*, February 17, 1907, 1.

26. Adrian Segura y Tornel, "Breves consideraciones acerca del tratamiento de las úlceras cutáneas" (MD thesis, Escuela de Medicina, 1874), 26.

27. Segura y Tornel, "Breves consideraciones," 28. Dr. John V. Goode of Dallas, Texas, also described the collodion process in detail: "Skin Grafting," *Annals of Surgery* 101, no. 3 (March 1935): 927–32.

28. French surgeon Louis de Wecker first developed the mosaic approach to skin grafts; "De la greffe dermique en chirurgie oculaire," *Annales d'oculistique* 68 (1872): 62.

29. Thomas Gibson, "Zoografting: A Curious Chapter in the History of Plastic Surgery," *British Journal of Plastic Surgery* 8 (1955): 234–42.

30. Luís Muñoz, "Cirugía práctica: sobre el ingerto epidérmico," *La Gaceta Médica de México* 5 (1870): 344–48.

31. M. Dubreuil, *Gazette des Hôpitaux*, July 30, 1872.

32. Thomas Bryant, "On Skin Grafting," *American Journal of the Medical Sciences* 64 (1872): 182.

33. Segura y Tornel, "Breves consideraciones," 34.

34. Anne Marie Moulin, "The Pasteur Institutes between the Two World Wars: The Transformation of the International Sanitary Order," in *Cambridge History of Medicine: International Health Organisations and Movements, 1918–1939*, ed. Paul Weindling (Cambridge: Cambridge University Press, 1995): 246; Ana Cecilia Rodríguez de Romo, "La ciencia pasteuriana a través de la vacuna antirrábica: el caso mexicano," *Dynamis* 16 (1996): 291–316; Eduardo Licéaga, *Las*

inoculaciones preventivas de la rabia (Mexico City: Imprenta de Ignacio Escalante, 1888).

35. Segura y Tornel, "Breves consideraciones," 36.

36. Licéaga, "Preamble to the Mexican Sanitary Code," 327.

37. Roy Porter, "The Patient's View: Doing Medical History from Below," *Theory and Society* 14, no. 2 (1985): 175–98.

38. Charles Rosenberg and Janet Golden, *Framing Disease: Studies in Cultural History* (New Brunswick, NJ: Rutgers University Press, 1992), xxiii.

39. Armus and López Denis, "Disease, Medicine, and Health," 428.

40. Birn and Necochea López, "Footprints on the Future," 509.

41. Trabulse, *Historia de la ciencia en México*, 16.

42. Dr. José Terrés, "Lecciones del Dr. José Terrés. Primera Lección," *Anales de la Escuela Nacional de Medicina. Parte Médica. Año I. 1904–1905* (Mexico City: Tipografía de los sucs. de Francisco Díaz de Leon, 1905), 14–42; Germán Fajardo-Dolci, Claudia Becerra Palars, Claudia Garrido, Eduardo de Ana Becerril, "El doctor José Terrés y su tiempo," *Historia de la medicina* 62, no. 3 (1999): 219–25.

43. "Escuela N. de Medicina: Primer Curso de Clinica Medica (Hospital Juárez)," 1908, AFM: FEMYA, leg. 190, exp. 10, fs. 156–57.

44. Alcohol intoxication significantly increased the risk of hurting or killing oneself, and doctors noted that some of the most serious injuries they treated resulted from drunkenness. John Bland-Sutton, "On Pulque and Pulque-Drinking in Mexico," *Lancet* 179, no. 4610 (1912): 45–46; Miguel Cicero, "Estudio anatomo-patológico del hígado de los enfermos alcohólicos" (MD thesis, Escuela Nacional de Medicina, 1872); Joaquín Herrera, "Algunas consideraciones relativas a la influencia del alcoholismo sobre la marcha de las heridas" (MD thesis, Facultad de Medicina, 1882), 10–11; "Segunda cuestión: del uso y del abuso de los alcohólicos," *Boletín del Consejo Superior de Salubridad del Distrito Federal* 3, no. 5–6 (December 31, 1882): 76–77.

45. Programa: propuesto por el profesor de 2° año de Clinica Externa para dar el curso durante el año escolar de 1895, December 21, 1894, AFM: FEMYA, leg. 192, exp. 1, fs. 81.

46. Rothschild, *Tratamiento de las quemaduras*, 15.

47. Guillermo Parra, "Algunas consideraciones sobre el hipnotismo desde el punto de vista terapéutico," *La Escuela de Medicina* 13, no. 21 (April 1, 1896): 463–66. Pierre Janet and Sigmund Freud used hypnosis to help sufferers of chronic pain, including burn victims; see Daniel E. Haycock, *Being and Perceiving* (New York: Manupod Press, 2011), 164.

48. Crescencio Garcia, "Armonicoterapia, ó sea la curación de las enfermedades por medio de la música," *La Escuela de Medicina* 9, no. 20 (1888): 416–54;

Juan Peon de Valle, "Sesión ordinaria del día 5 de Julio de 1901," *El Observador Médico* 1, no. 11 (1901): 173–76.

49. Gabriel González Olvera, "Vigésimaséptima Lección. Melesia Flores, soltera, de 24 años," *Anales de la Escuela Nacional de Medicina. Parte Médica. Año I. 1904–1905* (Mexico City: Tipografía de los sucs. de Francisco Díaz de Leon, 1905): 251–62.

50. "Higiene de la piel," *El Siglo Diez y Nueve*, November 5, 1874, 2.

51. González Olvera, "Vigésimaséptima Lección," 251–62.

52. Martínez Garza, "Apuntes de clínica externa," 45.

53. Claudia Agostoni, "Médicos científicos y medicos ilícitos en la Ciudad de México durante el Porfiriato," *Estudios de historia moderna y contemporánea de México* 19 (1999): 13–31.

54. "Crónica Médica—Los charlatans y curanderos," *La Gaceta Médica* 10, no. 1 (1875): 20–24.

55. Gamboa, *Santa*, 52.

56. Martínez Garza, "Apuntes de clínica externa," 45.

57. M. A. Posalagua, *Estudio para la formación de hospitales generales en la Ciudad de México* (Mexico City: Imprenta de Comercio de Nabor Chávez, 1874), 12.

58. J. Valenzuela, "De la asistencia médica a los enfermos pobres a domicilio," *La Escuela de Medicina* 3, no. 24 (1882): 329–31.

59. Valenzuela, "De la asistencia médica," 329–31.

60. González Urueña, "Descripción de una cura," 368–70.

61. Augusto López Amador, "Tratamiento de las quemaduras por la ambrina" (MD thesis, Facultad de Medicina UNAM, 1918), 24.

62. "Quemadura con azufre," *El Dictamen*, October 7, 1908, 4.

63. Takahiro Ueyama, *Health in the Marketplace: Professionalism, Therapeutic Desires, and Medical Commodification in Late-Victorian London* (Palo Alto, CA: Society for the Promotion of Science and Scholarship, 2010).

64. "Profesionales," *El Diario*, December 9, 1906, 8; "Profesionales," *El Diario*, December 10, 1906, 6; "Dr. J. E. McGarvin," *El Diario*, December 28, 1906, 3.

65. "Profesionales," *El Diario*, December 12, 1906, 8; "Profesionales," *El Diario*, December 14, 1906, 6.

66. "Conocimientos de gran utilidad para el hogar," *El Imparcial*, December 4, 1908, 4.

67. "Recetas Utiles," *El País*, August 31, 1907, 3; "Sección del doctor," *El Diario*, December 13, 1906, 6.

68. "Notas de Policia: Quemadures que producen la muerte," *El Imparcial*, January 27, 1905, 4.

69. "Remedio casero que envenena," *El Diario*, October 26, 1906, 2.

70. "El amigo de la salud: Ungüento Holloway," *El Siglo Diez y Nueve*, April 17, 1863, 4.

71. "El gran remedio Alemán," *El Siglo Diez y Nueve*, September 13, 1882, 3.

72. "El ungüento del Dr. Robinson," *El Siglo Diez y Nueve*, June 14, 1879, 4.

73. "Aporó," *El Siglo Diez y Nueve*, March 21, 1873, 4; "Balsamo Aporo," *El Siglo Diez y Nueve*, January 7, 1879, 6.

74. *La naturaleza: periódico científico de la sociedad Mexicana de historia natural*, 2nd series, vol. 2 (Mexico City: Imprenta de Ignacio Escalante, 1897), 258; "Excursión científica a Michoacán: Octubre de 1904. En compañía de los sres. Cyrus Pringle, George R. Show y Filomeno L. Lozano," *Anales del Instituto Médico Nacional* 7, no. 1 (Mexico City: Imprenta y Fototipía de la Secretaría de Fomento, 1905): 354–55.

75. Fernando Güereña, privilegio por la planta 'Raíz Güereña,' 1888, AGN: Patentes y Marcas, caja 36, exp. 1536.

76. Fernando Altamirano, "Breve Informe: Acerca de los trabajos hechos en el Instituto Médico Nacional para el estudio de la planta llamada 'Matarique' cacalia decomposita," *El Estudio: Semanario de Ciencias Médicas* 3, no. 6 (México, August 11, 1890): 81–83.

77. Altamirano, "Breve Informe," 81–83.

78. Nicolás León, *Biblioteca botánico-mexicana. Catálogo bibliográfico y crítico de autores y escritos referentes a vegetales de México y sus aplicaciones, desde la conquista hasta el presente* (Mexico City: Oficina Tipográfica de la Secretaría de Fomento, 1895), 13–14; *Botanisches Centralblatt: Referirendes Organ für Gesammtgebiet der Botanik des In und Auslandes* (Cassel: Verlag von Gebr. Gotthelft, 1890), 63.

79. Altamirano, "Breve Informe," 81–86.

80. Recently, several historians have unearthed similar stories of local or indigenous medical knowledge, and specifically, how authorities have appropriated the knowledge and profited from it. See Paul Gootenberg, *Andean Cocaine: The Making of a Global Drug* (Chapel Hill: University of North Carolina Press, 2008); Saul Jarcho, *Quinine's Predecessor: Francesco Torti and the Early History of Cinchona* (Baltimore, MD: Johns Hopkins University Press, 1993); Soto Laveaga, *Jungle Laboratories*.

CONCLUSION

1. National Board of Fire Underwriters, *Safeguarding Industry: A War-Time Necessity. Prepared for the Council of National Defense by the National Board of Fire Underwriters* (New York: National Board of Underwriters, 1917).

2. Knowles, *Disaster Experts*, 9.

3. Bird, *Protection against Fire*, 1–2.

4. "Corrido la quemazón," *Posada's Popular Mexican Prints*, 23.

5. Concejal Ingeniero Abraham Chávez to Secretario del Consejo Superior del Gobierno del Distrito, February 7, 1905, AHDF: AGDF, Consejo Superior de Gobierno del Distrito, diversiones pública, vol. 807, exp. 1323.

6. Bird, *Protection against Fire*, n.p.

7. Concejal Ingeniero Abraham Chávez to Secretario del Consejo Superior del Gobierno, August 7, 1912, AHDF: AGDF, Consejo Superior de Gobierno del Distrito, diversiones pública, vol. 596, exp. 19.

8. Concejal Ingeniero Abraham Chávez to Secretario del Consejo Superior del Gobierno, August 7, 1912, AHDF: AGDF, Consejo Superior de Gobierno del Distrito, diversiones pública, vol. 596, exp. 19.

SELECTED BIBLIOGRAPHY

ARCHIVES

Archivo General Municipal de Puebla (Puebla)- AGMP
Archivo General de la Nación (Mexico City)- AGN
    Comisión Monetaria
    Patentes y Marcas
    Segundo Imperio
    Suprema Corte de Justicia de la Nación- SCJN
    Tribunal Superior de Justicia del Distrito Federal- TSJDF
Archivo Histórico del Distrito Federal (Mexico City)- AHDF
    Ayuntamiento Gobierno del Distrito Federal- AGDF
    Carteles e Ilustraciones
    Gobierno del Distrito
    Planoteca
Archivo Histórico de Facultad de Medicina (Mexico City)- AFM
    La Biblioteca Dr. Nicolás León
    Fondo Escuela de Medicina y Alumnos- FEMYA
Archivo Histórico de la Secretaría de Salud (Mexico City)
Hagley Museum and Library (Wilmington, DE)
    Trade Catalogues
Universidad Nacional Autónoma de México (Mexico City)- UNAM
    Biblioteca Nacional
    Fondo Reservado de la Hemeroteca Nacional de México
    Archivo Histórico del UNAM, Fondo Escuela Nacional de Ingeniería
University of Hawaii at Manoa Library
    Jean Charlot Collection
University of Texas Libraries
    Perry-Castañeda Library Map Collection

SELECTED BIBLIOGRAPHY

Abbott, Andrew. *The System of Professions: An Essay on the Division of Expert Labor.* Chicago: University of Chicago Press, 1988.

Abu-Lughod, Lila, and Catherine A. Lutz. "Introduction: Emotion, Discourse, and the Politics of Everyday Life." In *Language and the Politics of Emotion,* edited by Catherine A. Lutz and Lila Abu-Lughod, 1–23. Cambridge: Cambridge University Press, 1990.

Adas, Michael. *Machines as the Measure of Men: Science, Technology, and Ideologies of Western Dominance.* Cornell Studies in Comparative History. Ithaca, NY: Cornell University Press, 1989.

Agostoni, Claudia. *Monuments of Progress: Modernization and Public Health in Mexico City, 1876–1910.* Boulder: University Press of Colorado, 2003.

Agostoni, Claudia. "Discurso médico, cultura higiénica y la mujer en la Ciudad de México al cambio de siglo (XIX–XX)." *Mexican Studies/Estudios Mexicanos* 18, no. 1 (2002): 1–22.

Agostoni, Claudia. "Médicos científicos y medicos ilícitos en la Ciudad de México durante el Porfiriato." *Estudios de historia moderna y contemporánea de México* no. 19 (1999): 13–31.

Aldrich, Mark. *Safety First: Technology, Labor, and Business in the Building of American Work Safety, 1870–1939.* Studies in Industry and Society. Baltimore, MD: Johns Hopkins University Press, 1997.

Almandoz, Arturo. "The Shaping of Venezuelan Urbanism in the Hygiene Debate of Caracas, 1880–1910." *Urban Studies* 37, no. 11 (2000): 2073–89.

Alonso, Angela Maria. *Idéias em movimento: a geração 1870 na crise do Brasil-Império.* São Paulo: ANPOCS, Paz e Terra, 2002.

Altamirano, Fernando. "Breve Informe: Acerca de los trabajos hechos en el Instituto Médico Nacional para el estudio de la planta llamada 'Matarique' cacalia decomposita." *El Estudio: Semanario de Ciencias Medicas* 3, no. 6 (México, August 11, 1890): 81–86.

Amador, José G. "'Redeeming the Tropics': Public Health and National Identity in Cuba, Puerto Rico, and Brazil, 1890–1940." PhD diss., University of Michigan, 2008.

American School of Correspondence. *Cyclopedia of Fire Prevention and Insurance: A General Reference Work, Prepared by Architects, Engineers, and Practical Insurance Men.* Vol. 1. Chicago: American School of Correspondence, 1912.

Anales del Ministerio de Fomento de la República Mexicana, Tomo IV. México: Imprenta de Francisco Díaz de Leon, 1881.

Andrade, Manuel. "Nuevo tratamiento de la quemaduras." *La Gaceta Médica de México* 3 (1838): 347–52.

Aréchiga Córdoba, Ernesto. "La ciudad y Tepito: a la zaga de la higiene pública." In *Tepito: del antiguo barrio de indios al arrabal*, edited by Ernesto Aréchiga Córdoba, 171–207. Mexico City: UNÍOS, 2003.

Arellano, J. J. R. de, and D. Orvañanos. "Consejo Superior de Salubridad." *El Estudio: Semanario de Ciencias Medicas* 1, no. 27 (December 9, 1889): 431–32.

Armus, Diego. *The Ailing City: Health, Tuberculosis, and Culture in Buenos Aires, 1870–1950*. Durham, NC: Duke University Press, 2011.

Armus, Diego, and Adrián López Denis. "Disease, Medicine, and Health." In *The Oxford Handbook of Latin American History*, edited by José C. Moya, 424–53. Oxford: Oxford University Press, 2010.

Arrom, Silvia Marina. *Containing the Poor: The Mexico City Poor House, 1775–1871*. Durham, NC: Duke University Press, 2000.

Arróniz, Marcos. *Manual del viajero en México*. Mexico City: Instituto Mora, 1991. First published 1858 by Librería de Rosa y Bouret.

Auyero, Javier, and Débora Alejandra Swistun. *Flammable: Environmental Suffering in an Argentine Shantytown*. Oxford: Oxford University Press, 2009.

Bankoff, Greg. *Cultures of Disaster: Society and Natural Hazard in the Philippines*. London: RoutledgeCurzon, 2003.

Bannister, Jon, and Nick Fyfe. "Introduction: Fear and the City." *Urban Studies* 38, no. 5–6 (2001): 807–13.

Baranoff, Dalit. "Shaped by Risk: The American Fire Insurance Industry, 1790–1920." PhD diss., Johns Hopkins University, 2004.

Barcena, Mariano. *La 2. Exposición de 'las clases productoras,' y descripción de la ciudad de Guadalajara*. Guadalajara: Tipografía de Sinforoso Banda, 1880.

Baynton, Thomas. *Descriptive Account of a New Method of Treating Old Ulcers of the Legs*. 2nd ed. London: Emery and Adams, 1799.

Beatty, Edward. "Approaches to Technology Transfer in History and the Case of Nineteenth-Century Mexico." *Comparative Technology Transfer and Society* 1, no. 2 (2003): 167–97.

Beatty, Edward. "Patents and Technological Change in Late Industrialization: Nineteenth-Century Mexico in Comparative Context." *History of Technology* 24 (2002): 121–50.

Beatty, Edward N., and Lucrecia Orensanz Escofet. "Invención e innovación: ley de patentes y tecnología en el México del siglo XIX." *Historia Mexicana* 45, no. 3 (1996): 567–619.

Beck, Ulrich. *Risk Society: Towards a New Modernity*. London: Sage, 1992.

Beezley, William H. *Judas at the Jockey Club and Other Episodes of Porfirian Mexico*. 2nd. ed. Lincoln: University of Nebraska Press, 2004.

Beezley, William H. "Mexican Sartre on the Zócalo: Nicolás Zúñiga y Miranda." In *The Human Tradition in Latin America: The Nineteenth Century*, edited by Judith Ewell and William H. Beezley, 204–14. Wilmington, DE: Scholarly Resources, 1989.

Benjamin, Thomas, and Marcial Ocasio-Meléndez. "Organizing the Memory of Modern Mexico: Porfirian Historiography in Perspective, 1880s–1980s." *Hispanic American Historical Review* 64, no. 2 (1984): 323–64.

Berdecio, Roberto, and Stanley Appelbaum, eds. *Posada's Popular Mexican Prints: 273 Cuts by José Guadalupe Posada*. New York: Dover, 1972.

Berra Stoppa, Erica. "La expansión de la Ciudad de México y los conflictos urbanos, 1900–1930." PhD diss., Colegio de México, 1983.

Bijker, Wiebe E., Thomas P. Hughes, and Trevor Pinch. *The Social Construction of Technological Systems: New Directions in the Sociology and History of Technology*. Cambridge, MA: MIT Press, 1989.

Bird, Joseph. *Protection against Fire, and the Best Means of Putting out Fires in Cities, Towns, and Villages, with Practical Suggestions for the Security of Life and Property*. New York: Hurd and Houghton, 1873.

Birn, Anne-Emanuelle, and Raúl Necochea López. "Footprints on the Future: Looking Forward to the History of Health and Medicine in Latin America in the Twenty-First Century." *HAHR* 91, no. 3 (2011): 503–27.

Bland-Sutton, John. "On Pulque and Pulque-Drinking in Mexico." *Lancet* 179, no. 4610 (1912): 43–46.

Blasio, José Luis. *Maximilian, Emperor of Mexico: Memoirs of His Private Secretary*. Translated by Robert Hammond Murray. New Haven, CT: Yale University Press, 1934.

Blum, Ann S. "Conspicuous Benevolence: Liberalism, Public Welfare, and Private Charity in Porfirian Mexico City, 1877–1910." *Americas* 58, no. 1 (2001): 7–38.

Boletín de las leyes del Imperio Mexicano ó sea codigo de la restauración. Vol. 4. Mexico City: José Sebastian Segura, 1865.

Bolton Valenčius, Conevery. *The Health of the Country: How American Settlers Understood Themselves and Their Land*. New York: Basic Books, 2002.

Botanisches Centralblatt: Referirendes Organ für Gesammtgebiet der Botanik des In- und Auslandes. Cassel: Verlag von Gebr. Gotthelft, 1890.

Bourke, Joanna. "Fear and Anxiety: Writing about Emotion in Modern History." *History Workshop Journal* 55, no. 1 (2003): 111–33.

Boyer, Christopher R. *Political Landscapes: Forest, Conservation, and Community in Mexico*. Durham, NC: Duke University Press, 2015.

Bribiesca, Juan. *Memoria documentada de los trabajos municipales de 1900*. Vol. 1. Mexico City: La Europea, 1901.

Bryant, Thomas. "On Skin Grafting." *American Journal of the Medical Sciences* 64 (1872): 182.

Bunker, Steven B. *Creating Mexican Consumer Culture in the Age of Porfirio Díaz*. Albuquerque: University of New Mexico Press, 2012.

Calderón de la Barca, Fanny. *Life in Mexico: The Letters of Fanny Calderón de la Barca with New Material from the Author's Private Journals*. Edited by Howard T. Fisher and Marion Hall Fisher. Garden City, NY: Doubleday, 1966.

Calvert, Monte A. *The Mechanical Engineer in America, 1830–1910: Professional Cultures in Conflict*. Baltimore, MD: Johns Hopkins University Press, 1967.

Carey, Mark. *In the Shadow of Melting Glaciers: Climate Change and Andean Society*. Oxford: Oxford University Press, 2010.

Carson, Rachel. *Silent Spring*. Boston: Houghton Mifflin Company, 1962.

Castillo Velasco, José M. del. *Colección de leyes, supremas órdenes, bandos, disposiciones de policía y reglamentos municipals de administración del Distrito Federal*. 2nd ed. Mexico City: Impreso por Castillo Velasco é Hijos, 1874.

Cerutti, Mario. *Empresarios españoles y sociedad capitalista en México (1840–1920)*. Asturias: Fundación Archivo de Indianos, 1995.

Chalhoub, Sidney. "The Politics of Disease Control: Yellow Fever and Race in Nineteenth-Century Rio de Janeiro." *Journal of Latin American Studies* 25, no. 3 (1993): 441–63.

Champlin, Henry L. *The American Firemen: Essays, Lurid Leaves, Sketches, Sparks: A Standard Work on Fire Matters*. Boston: H. L. Champlin, 1875.

Chávez Franco, Modesto. *Historia general del cuerpo de bomberos de Guayaquil*. 2nd ed. Guayaquil: Banco Central del Ecuador, 1985.

Choquet, Édouard. *Les Incendies dans les Théâtres*. Paris: Baudry, 1886.

Cicero, Miguel. "Estudio anatomo-patológico del hígado de los enfermos alcohólicos." MD thesis, Escuela Nacional de Medicina, 1872.

Clark, Claudia. *Radium Girls: Women and Industrial Health Reform, 1910–1935*. Chapel Hill: University of North Carolina Press, 1997.

Clark de Lara, Belem, and Mariana Flores Monroy. *El renacimiento: periódico literario, segunda época*. Mexico City: Universidad Autónoma de México, 2006.

Coffin, Alfred Oscar. *Land without Chimneys; or, The Byways of Mexico*. Cincinnati: Editor Publishing, 1898.

Conkling, Howard. *Mexico and the Mexicans: Or, Notes of Travel in the Winter and Spring of 1883*. New York: Taintor Brothers, Merrill and Co., 1883.

Corbin, Alain. *El perfume o el miasma: el olfato y lo imaginario social, siglos XVIII y XIX*. Sección de obras de historia. Mexico City: Fondo de Cultura Económica, 1987.

Corn, Joseph J. *User Unfriendly: Consumer Struggles with Personal Technologies, from Clocks and Sewing Machines to Cars and Computers*. Baltimore, MD: Johns Hopkins University Press, 2011.

Crespo Martínez, Gilberto. *Las patentes de invención*. Mexico City: Oficina Tipografía de la Secretaría de Fomento, 1897.

"Cuarta Comision de Fábricas e Industrias." *Boletín del Consejo de Salubridad del Distrito Federal* 2, no. 9 (March 31, 1882): 138–39.

Cuchí Espada, Víctor. "Cambio de costumbres o cómo ser moderno: comerciantes, Ayuntamiento y mercado telefónico en la Ciudad de México, 1881–1905." *Cuicuilco* 15 (1999): 265–303.

Cutter, Susan. "Vulnerability to Environmental Hazards." *Progress in Human Geography* 20, no. 4 (1996): 529–39.

David, Paul A. *Technological Choice Innovation and Economic Growth: Essays on American and British Experience in the Nineteenth Century*. Cambridge: Cambridge University Press, 1975.

Dean, A. F. "Fire Insurance." In *Lectures on Commerce Delivered before the College of Commerce and Administration of the University of Chicago*, edited by Henry Rand Hatfield, 321–79. Chicago: University of Chicago Press, 1907.

Dean, Warren. *The Industrialization of São Paulo, 1880–1945*. Austin: University of Texas Press, 1969.

Delobel (de Noyon), Julio. "Higiene del escolar." *Boletín del Consejo Superior de Salubridad 3a Época* 7, no. 1 (June 31, 1901): 1–19.

Demeritt, David. "Ecology, Objectivity, and Critique in Writings on Nature and Human Societies." *Journal of Historical Geography* 20, no. 1 (1994): 22–37.

Díaz Rugama, Adolfo. "Distribución y legislación de aguas en las ciudades." In *Concurso Científico. Asociación de Ingenieros y Arquitectos. Discurso pronunciado en la sesión del 22 de julio de 1895*. Mexico City: Oficina Tipografía de la Secretaría de Fomento, 1895.

Dios Peza, Juan de. *La beneficencia en México*. Mexico City: Imprenta de Francisco Díaz de León, 1881.

Dublán, Manuel, and José María Lozano. *Legislación Mexicana: Colección completa de las disposiciones legislativas expedidas desde la independencia de la República*. Vol. 12. Mexico City: Imprenta y Litografía de Eduardo Dublán y Comp., 1886.

Ducat, Arthur C. *The Practice of Fire Underwriting*. 4th ed. New York: T. Johns Jr., 1866.

Durham, Alan L. *Patent Law Essentials: A Concise Guide*. 2nd ed. Westport, CT: Praeger, 2004.

Ellis, Oliver C. *A History of Fire and Flame 1932*. London: J. W. Northend, 1932.

Enciso Ulloa, Antonio, H. G. Perkins, and Louis C. Simonds, *The Mexican Pocket Lawyer*. Mexico City: Imprenta de Hull, 1910.

Endfield, Georgina H. *Climate and Society in Colonial Mexico: A Study of Vulnerability*. Malden, MA: Blackwell Publishing, 2008.

Enticknap, Leo. "The Film Industry's Conversion from Nitrate to Safety Film in the Late 1940s: A Discussion of the Reasons and Consequences." In *This Film Is Dangerous: A Celebration of Nitrate Film*, edited by Roger Smither and Catherine Surowiec, 202–12. Brussels: FIAF, 2002.

Espinosa, Mariola. "Globalizing the History of Disease, Medicine, and Public Health in Latin America." *Isis* 104, no. 4 (2013): 798–806.

Evans, Peter. *Dependent Development: The Alliance of Multinational, State and Local Capital in Brazil*. Princeton, NJ: Princeton University Press, 1979.

"Excursión científica a Michoacán: Octubre de 1904. En compañía de los sres. Cyrus Pringle, George R. Show y Filomeno L. Lozano." *Anales del Instituto Médico Nacional* 7, no. 1, 341–55. Mexico City: Imprenta y Fototipía de la Secretaría de Fomento, 1905.

"Expediente: Relativo a mejoras introducidas en la fabricación de los cerillos por los Sres. Lascurain y Compañía." *Boletín del Consejo de Salubridad del Distrito Federal* 3, no. 3–4 (1882): 41–42.

Fajardo-Dolci, Germán, Claudia Becerra Palars, Claudia Garrido, and Eduardo de Ana Becerril. "El doctor José Terrés y su tiempo." *Historia de la medicina* 62, no. 3 (1999): 219–25.

Febvre, Lucien. "Sensibility and History: How to Reconstitute the Emotional Life of the Past." In *A New Kind of History: From the Writings of Febvre*, edited by Peter Burke, translated by K. Folca, 12–26. London: Routledge and Kegan Paul, 1973.

Fenn, Elizabeth A. *Pox Americana: The Great Smallpox Epidemic of 1775–82*. New York: Hill and Wang, 2001.

Fernández-Christlieb, Federico. *Mexico, Ville Néoclassique: Les espaces et les idées d'aménagement urbain (1783–1911)*. Paris: L'Harmattan, 2002.

Figueroa Doménech, J. *Guía general descriptive de la República Mexicana*. Vol. 2. Barcelona: R. de S.N. Araluce, 1899.

Fisher, John, and Natalie Priego. "Ignorance and 'Habitus': Blinkered and Enlightened Approaches towards the History of Science in Latin America." *Bulletin of Latin American Research* 25, no. 4 (2006): 528–40.

Foelsch, August. *Theaterbrände und die zur Verhütung derselben erforderlichen Schutz-Massregelu; mit einem Verzeichniss von 523 abgebrannten Theatern*. Hamburg: O. Meissner, 1878.

Fowler, J. A. *The Fire Account—Physical, Personal, Moral. Address Delivered before the Fire Underwriters' Association of the Northwest*. Chicago, 1878.

Frank, Patrick. *Posada's Broadsheets: Mexican Popular Imagery, 1890–1910*. Albuquerque: University of New Mexico Press, 1998.

Frazier, W. H. *Few and Furious: Also, a Reprint of the Second Edition of the Little Paragraphs: Comments and Criticisms on the Business of Fire Insurance*. Philadelphia, 1902.

Freeman, Christopher. *Technology Policy and Economic Performance: Lessons from Japan*. London: Pinter, 1987.

French, William E. "Imagining and the Cultural History of Nineteenth-Century Mexico." *Hispanic American Historical Review* 79, no. 2 (1999): 249–67.

Freshwater, M. Felix, and Thomas J. Krizek. "George David Pollock and the Development of Skin Grafting." *Annals of Plastic Surgery* 1, no. 1 (1978): 96–104.

Galindo y Villa, Jesús. *Reseña histórica-descriptiva de la Ciudad de México que escribe el Regidor del Ayuntamiento, por encargo del Señor Presidente de la misma corporación D. Guillermo Landa y Escandón, y expresamente para los delegados a la segunda conferencia internacional americana*. Mexico City: Díaz de León, 1901.

Gamboa, Federico. *Santa: A Novel of Mexico City*. Translated by John Charles Chasteen. Chapel Hill: University of North Carolina Press, 2010.

Garay, Francisco de. *Discurso pronunciado por el ingeniero Francisco de Garay en la Asociación de Ingenieros Civiles y Arquitectos, al tomar posesión de la presidencia de la misma*. Mexico City: Imprenta del Comercio de Dublán y Chávez, 1877.

García, Crescencio. "Armonicoterapia, ó sea la curación de las enfermedades por medio de la música." *La Escuela de Medicina* 9, no. 20 (October 31, 1888): 416–54.

García Acosta, Virginia, and Gerardo Suarez Reyñoso. *Los sismos en la historia de Mexico*. Vol. 1. Mexico City: Universidad Nacional Autónoma de México, 1996.

García Aguirre, Feliciano J., and Emilia Valdez Méndez. "Dos Bocas: una contribucion a la historia de los desastres en Veracruz." *Instituto de Investigaciones Historico-Sociales* (1995): 105–21.

Garza, James. *The Imagined Underworld: Sex, Crime, and Vice in Porfirian Mexico City*. Lincoln: University of Nebraska Press, 2007.

Garza Merodio, Gustavo G. "Technological Innovation and the Expansion of Mexico City, 1870–1920." *Journal of Latin American Geography* 5, no. 2 (2006): 109–26.

García Sepúlveda, Federico. "Estadística general del Hospital Juárez, 1888–1895." MD thesis, Escuela de Medicina, 1896.

Garmendia, José María. *República Mexicana, Secretaría de Estado y del Despacho de Hacienda y Crédito Pública: Noticia de la importación y exportación de mercancías en los años fiscales de 1872 á 1873, 1873 á 1874 y 1874 á 1875*. Mexico City: Tipografía de Gonzalo A. Esteva, 1880.

Gayol, Roberto. "Reflexiones sugeridas por el Art. 257 del Codigo Sanitario que se refiere a las obras públicas que interesan a la higiene." *Boletín del Consejo Superior de Salubridad del Distrito Federal 3a Época* 3, no. 5 (1897): 135–55.

Gerhard, William Paul. *Theatre Fires and Panics: Their Causes and Prevention.* New York: John Wiley and Sons, 1896.

Gerster, Arpad G. *The Rules of Aseptic and Antiseptic Surgery: A Practical Treatise for the Use of Students and the General Practitioner.* New York: D. Appleton and Company, 1890.

Gibson, Thomas. "Zoografting: A Curious Chapter in the History of Plastic Surgery." *British Journal of Plastic Surgery* 8, no. 3 (1955): 234–42.

Gilliland, Jason A., and Mathew Novak. "On Positioning the Past with the Present: The Use of Fire Insurance Plans and GIS for Urban Environmental History." *Environmental History* 11, no. 1 (2006): 136–39.

González Navarro, Moisés. "Illness and Mortality." In *The Age of Porfirio Díaz, Selected Readings,* edited by Carlos B. Gil, 111–15. Albuquerque: University of New Mexico Press, 1977.

González Navarro, Moisés. *La pobreza en México.* Mexico City: Colegio de México, 1985.

González Olvera, Gabriel. "Vigésimaséptima Lección. Melesia Flores, soltera, de 24 años." *Anales de la Escuela Nacional de Medicina. Parte Médica. Año I. 1904–1905,* 251–62. Mexico City: Tipografía de los sucs. de Francisco Díaz de Leon, 1905.

González Urueña, Juan Manuel. "Descripción de una cura de quemadura causada por la inflamación de los gases que se desprenden de las letrinas." *La Gaceta Médica de México* 1, no. 368 (1836): 368–70.

Goode, John V. "Skin Grafting." *Annals of Surgery* 101, no. 3 (1935): 927–32.

Greenberg, Amy S. *Cause for Alarm: The Volunteer Fire Department in the Nineteenth-Century City.* Princeton, NJ: Princeton University Press, 1998.

Greenberg, Amy S. "The Origins of the Municipal Fire Department: Nineteenth-Century Change from an International Perspective." In *Municipal Services and Employees in the Modern City: New Historical Approaches,* edited by Michèle Dagenais, Irene Maver, and Pierre-Yves Saunier, 47–65. Aldershot, UK: Ashgate, 2003.

Greenhow, Edward. "New Method of Treating Burns." *London Medical Gazette* 1 (1839): 82–84.

Guerra, François-Xavier. *México: del antiguo régimen a la revolución.* Mexico City: Fondo de Cultura Económica, 1988.

Halbwachs, Maurice. *On Collective Memory.* Chicago: University of Chicago Press, 1992.

Hardoy, Jorge E. "Theory and Practice of Urban Planning in Europe, 1850–1930." In *Rethinking the Latin American City,* edited by Richard M. Morse and Jorge E. Hardoy, 20–49. Washington, DC: Woodrow Wilson Center Press, 1992.

Harris, Robert L. "PCIH Presentation: The Public Health Roots of Industrial Hygiene." *American Industrial Hygiene Association Journal* 58, no. 3 (1997): 176–79.

Haycock, Daniel E. *Being and Perceiving*. New York: Manupod Press, 2011.

Hayward, R. J. "Insurance Plans and Land Use Atlases: Sources for Urban Historical Research." *Urban History Review* 2 (1973): 2–9.

Heckman, Heather. "Burn after Viewing, or, Fire in the Vaults: Nitrate Decomposition and Combustibility." *American Archivist* 73, no. 2 (2010): 483–506.

Herrera, Joaquín. "Algunas consideraciones relativas a la influencia del alcoholismo sobre la marcha de las heridas." MD thesis, Facultad de Medicina, 1882.

Hexamer, Charles John. *Fire Hazards in Textile Mills, Mill Architecture, and Means for Extinguishing Fire: Three Lectures Delivered before the Franklin Institute*. Philadelphia: Merrihew Print, J. Spencer Smith, 1885.

Hexamer, Charles John. *On the Prevention of Fires in Theatres*. Philadelphia: Merrihew Print, 1882.

Hill, Libby. *The Chicago River: A Natural and Unnatural History*. Chicago: Lake Claremont Press, 2000.

Hoffman, Richard C., Nancy Langston, James McCann, Peter Perdue, and Lise Sedrez. "AHR Conversation: Environmental Historians and Environmental Crisis." *American Historical Review* 113, no. 5 (2008): 1431–65.

Hounshell, David. "Rethinking the History of 'American Technology.'" In *In Context: History and the History of Technology, Essays in Honor of Melvin Kranzberg*, edited by Stephen H. Cutcliffe and Robert C. Post, 216–29. Bethlehem, PA: Lehigh University Press, 1989.

Hounshell, David A., and John Kenly Smith, Jr. *Science and Corporate Strategy: Du Pont R&D, 1902–1980*. Studies in Economic History and Policy: The United States in the Twentieth Century. New York: Cambridge University Press, 1988.

Hughes, Thomas P. *American Genesis: A Century of Invention and Technological Enthusiasm, 1870–1970*. Chicago: University of Chicago Press, 1989.

Hughes, Thomas P. *Elmer Sperry: Inventor and Engineer*. Baltimore, MD: Johns Hopkins Press, 1971.

Ibarrola, José Ramón de. *Apuntes sobre el desarrollo de la ingeniería en México y la educación del ingeniero*. Mexico City: Tipografía de la viuda de F. Díaz de León, 1911.

"Informe: Relativo a las condiciones higiénicas que deben satisfacer los teatros y otras salas de espectáculo." *Boletín del Consejo de Salubridad del Distrito Federal* 3, no. 7–8 (February 28, 1883): 100–101.

"Iniciativa: para la reglamentación en el Distrito Federal de los establecimientos peligrosos, insalubres, e incomodos." *Boletín del Consejo Superior de Salubridad* 3, no. 1–2 (August 31, 1882): 1–3.

Institution of Civil Engineers. *Power for the Use of Man: A Series of Addresses to the General Assembly of the Institution of Civil Engineers, Commemorating the 150th Anniversary of the Royal Charter of the Institution granted on 3 June 1828.* London: The Institution of Civil Engineers, 1978.

Janis, Mark D. "Patent Abolitionism." *Berkeley Technology Law Journal* 17, no. 2 (2002): 899–952.

Jefferson, Mark. "The Culture of the Nations." *Bulletin of the American Geographical Society* 43, no. 4 (1911): 256.

Jefferson, Mark. "The Geographic Distribution of Inventiveness." *Geographical Review* 19, no. 4 (1929): 649–61.

Jenkins, D. T. "The Practice and Insurance against Fire, 1750–1840, and Historical Research." In *The Historian and the Business of Insurance*, edited by Oliver M. Westall, 18–19. Manchester: Manchester University Press, 1984.

Jiménez, Christina. "Popular Organizing for Public Services: Residents Modernize Morelia, Mexico, 1880–1920." *Journal of Urban History* 30, no. 4 (2004): 495–518.

Johns, Michael. *The City of Mexico in the Age of Díaz.* Austin: University of Texas Press, 1997.

Jones, Gareth. "The Latin American City as Contested Space: A Manifesto." *Bulletin of Latin American Research* 13, no. 1 (1994): 1–12.

Josephson, Paul R. *Industrialized Nature: Brute Force Technology and the Transformation of the Natural World.* Washington, DC: Island Press/Shearwater Books, 2002.

Juuti, Petri S., Tapio S. Katko, and Heikki S. Vuorinen. *Environmental History of Water: Global Views on Community Water Supply and Sanitation.* London: IWA Publishing, 2007.

Kargon, Robert H., and Scott G. Knowles. "Knowledge for Use: Science, Higher Learning, and America's New Industrial Heartland, 1880–1915." *Annals of Science* 59 (2002): 1–20.

Kenlon, John. *Fires and Fire-Fighters: A History of Modern Fire-Fighting with a Review of Its Development from Earliest Times.* New York: George H. Doran Company, 1913.

Kenny, Daniel J. *Cincinnati Exposition Guide and Catalogue of the Fine Arts Department, Containing the Name and Address of Every Exhibitor at the Fifth Cincinnati Industrial Exposition of 1874.* Cincinnati, OH: Cincinnati Gazette Co., 1874.

Keyes, Sarah. "'Like a Roaring Lion': The Overland Trail as a Sonic Conquest." *Journal of American History* 96, no. 1 (2009): 19–43.

Klein, Naomi. *The Shock Doctrine: The Rise of Disaster Capitalism.* New York: Metropolitan Books, 2007.

Knowles, Scott Gabriel. *The Disaster Experts: Mastering Risk in Modern America.* Philadelphia: University of Pennsylvania Press, 2011.

Krafft, Thomas. "Reconstructing the North American Urban Landscape: Fire Insurance Maps—An Indispensable Source (Feuerversicherungskarten: Unverzichtbares Hilfsmittel zur historisch—geographischen Rekonstruktion der nordamerikanischen Stadt)." *Erdukunde* 47, no. 3 (1993): 196–211.

Kuecker, Glen D. "'The Greatest and the Worst': Dominant and Subaltern Memories of the Dos Bocas Well Fire of 1908." In *The Memory of Catastrophe,* edited by Peter Gray and Kendrick Oliver, 65–78. Manchester: Manchester University Press, 2004.

Leal, Juan Félipe. *Del mutualismo al sindicalismo: 1843–1910.* Mexico City: Ediciones El Caballito, 1991.

Leal, Juan Félipe, and Eduardo Barraza. "Inicios de la reglamentación cinematográfica en la Ciudad de México." *Revista Mexicana de Ciencias Políticas y Sociales* no. 150 (1992): 139–77.

Leal, Juan Félipe, Eduardo Barraza, and Carlos Flores. *El arcón de las vistas: cartelera del cine en México, 1896–1910.* Mexico City: Universidad Nacional Autónoma de México, 1994.

Lear, John. *Workers, Neighbors, and Citizens: The Revolution in Mexico City.* Lincoln: University of Nebraska Press, 2001.

Lefebvre, Henri. "Reflections on the Politics of Space." Translated by Michael J. Enders. *Antipode* 8, no. 2 (1976): 30–37.

León, Nicolás. *Biblioteca botánico-mexicana. Catálogo bibliográfico y crítico de autores y escritos referentes a vegetales de México y sus aplicaciones, desde la conquista hasta el presente.* Mexico City: Oficina Tipográfica de la Secretaría de Fomento, 1895.

Licéaga, Eduardo. "Address." *Public Health Papers and Reports* 18 (1892): 15–24.

Licéaga, Eduardo. *Las inoculaciones preventivas de la rabia.* Mexico City: Imprenta de Ignacio Escalante, 1888.

Licéaga, Eduardo. "Preamble to the Mexican Sanitary Code." In *Historia de la Salubridad y de la Asistencia en México.* Vol. 1. Edited by José Alvarez Amézquita et al. Mexico City: Secretaría de Salubridad y Asistencia, 1960.

Licht, Walter. *Working for the Railroad: The Organization of Work in the Nineteenth Century.* Princeton, NJ: Princeton University Press, 1987.

Lira González, Andrés. *Comunidades indígenas frente a la Ciudad de México. Tenochtitlán y Tlatelolco, sus pueblos y barrios, 1812–1919.* Zamora, Michoacán: Colegio de Michoacán-Colegio de Mexico-Consejo Nacional de Ciencia y Tecnología, 1983.

López Amador, Augusto. "Tratamiento de las quemaduras por la ambrina." MD thesis, Facultad de Medicina Universidad Autónoma de México, 1918.

Lübken, Uwe, and Christof Mauch. "Uncertain Environments: Natural Hazards, Risk, and Insurance in Historical Perspective." *Environment and History* 17, no. 1 (2011): 1–12.

Lussier, Hubert. *Les Sapeurs-Pompiers au XIXe siècle: Associations volontaires en milieu populaire.* Paris: L'Harmattan/Association des Ruralistes Français, 1987.

Mackenbach, Johan P. "Politics Is Nothing but Medicine at a Larger Scale: Reflections on Public Health's Biggest Idea." *Journal of Epidemiology and Community Health* 63, no. 3 (2009): 181–84.

MacKenzie, Donald. *Inventing Accuracy: A Historical Sociology of Nuclear Missile Guidance.* Cambridge, MA: MIT Press, 1990.

MacLeod, Christine. "Concepts of Invention and the Patent Controversy in Victorian Britain." In *Technological Change: Methods and Themes in the History of Technology*, edited by Robert Fox, 137–53. Amsterdam: Harwood Academic, 1996.

MacLeod, Christine. "The Invention of Heroes." *Nature* 460, no. 30 (2009): 572–73.

Mallon, Bill. *The 1900 Olympic Games: Results for All Competitors in All Events, with Commentary.* Jefferson, NC: McFarland, 1998.

Márquez, Patricio V., and Daniel J. Joly. "A Historical Overview of the Ministries of Public Health and Medical Programs of the Social Security Systems in Latin America." *Journal of Public Health Policy* 7, no. 3 (1986): 378–94.

The Massey-Gilbert Blue Book of Mexico: A Directory in English of Mexico City. International Bank and Trust Company of America, 1903.

Martínez Garza, Pedro. "Apuntes de clínica externa." MD thesis, Escuela de Medicina, 1895.

Mauelshagen, Franz. "Disaster and Political Culture in Germany since 1500." In *Natural Disasters, Cultural Responses: Case Studies toward a Global Environmental History*, edited by Christof Mauch and Christian Pfister, 41–76. Lanham, MD: Lexington Books, 2009.

McEvoy, Arthur F. "The Triangle Shirtwaist Factory Fire of 1911: Social Change, Industrial Accidents, and the Evolution of Common-Sense Causality." *Law and Social Inquiry* 20, no. 2 (1995): 621–51.

McEvoy, Arthur F. "Working Environments: An Ecological Approach to Industrial Health and Safety." *Technology and Culture* 36, no. 2 (1995): S145–73.

Medina, Carlos A. de. *Exposición que hace el Ingeniero Carlos A. de Medina a todos los habitantes de la Ciudad de México sobre las grandes ventajas que trae consigo para la capital.* Mexico City: Imprenta y Litografía de Dublan y Ca., 1884.

Meléndez Marín, Samuel. *Tijuana crece al calor de las llamas*. Tijuana: Editorial Zenit, 1983.

Melosi, Martin V. *The Sanitary City: Urban Infrastructure in America from Colonial Times to the Present*. Baltimore, MD: Johns Hopkins University Press, 2000.

Memoria de la Exposición Municipal de 1874. México: Imprenta de Díaz de Leon y White, 1874.

Memoria presentada á S.M. el Emperador por el Ministro de Fomento Luis Robles Pezuela de los trabajos ejecutados en su ramo el año de 1865. México: Imprenta de J. M. Andrade y F. Escalante, 1866.

Mendez, Luis. *Gaceta de los Tribunales de la República Mexicana*. Vol. 3. México: Isidoro Devaux, 1862.

Ministère du Commerce, de L'Industrie des postes et des télégraphes. *Exposition Universelle Internationale de 1900. Congrès international das sapeurs-pompiers: tenu à Paris le 12 août 1900: procès-verbal sommaire*. Paris: Imprimerie nationale, 1901.

Minzoni Consorti, Antonio. *Crónica de dos siglos de seguro en México*. Mexico City: Comisión Nacional de Seguros y Fianzas, 2005.

Moore, F. C. *Fires: Their Causes, Prevention and Extinction, Combining Also a Guide to Agents Respecting Insurance against Loss by Fire*. New York: Continental Insurance Co., 1877.

Moulin, Anne Marie. "The Pasteur Institutes between the Two World Wars: The Transformation of the International Sanitary Order." In *Cambridge History of Medicine: International Health Organisations and Movements, 1918–1939*, edited by Paul Weindling, 244–65. Cambridge: Cambridge University Press, 1995.

Mumford, Lewis. "What Is a City?" *Architectural Record* 82, no. 5 (1937): 92–95.

Muñoz, Luís. "Cirugía práctica: sobre el ingerto epidérmico." *La Gaceta Médica de México* 5 (1870): 344–48.

Myers, Melvin, and James D. McGlothlin. "Matchmakers' 'Phossy Jaw' Eradicated." *American Industrial Hygiene Association Journal* 57, no. 4 (1996): 330–32.

Nash, Linda. *Inescapable Ecologies: A History of Environment, Disease, and Knowledge*. Berkeley: University of California Press, 2006.

National Board of Fire Underwriters. *Fire Insurance: Its Importance; Its Relation to the Community*. New York: Press of Styles and Cash, 1899.

National Board of Fire Underwriters. *Safeguarding Industry: A War-Time Necessity. Prepared for the Council of National Defense by the National Board of Fire Underwriters*. New York: National Board of Fire Underwriters, 1917.

La naturaleza: periódico científico de la sociedad Mexicana de historia natural. 2nd series. Vol. 1. Mexico City: Imprenta de Ignacio Escalante, 1897.

Neefus, H. F. *Sprinkler Equipments: Installation and Requirements*. Newark, NJ: The Merchants' Insurance Co., 1895.

Neill, Deborah. *Networks in Tropical Medicine: Internationalism, Colonialism, and the Rise of a Medical Specialty, 1890–1930*. Stanford: Stanford University Press, 2012.

Nélaton, Auguste. *Élémens de pathologie chirurgicale*. Paris: Germer Bailliére, Libraire-Éditeur, 1844.

Nelson, Richard R. *National Innovation Systems: A Comparative Analysis*. Oxford: Oxford University Press, 1993.

Neufeld, Stephen. "Servants of the Nation: The Military in the Making of Modern Mexico, 1876–1911." PhD diss., University of Arizona, 2009.

Oliveira Sobrinho, Afonso Soares de. "São Paulo e a Ideologia Higienista entre os séculos XIX e XX a utopia de civilidade." *Sociologia* 15, no. 32 (2013): 210–35.

Overland Monthly: The Awakening of Mexico Centenary of the Republic. Vol. 56 (July 1910).

Pacheco Salazar, Victor F., and Héctor L. Ocaña Servín. "La legislación ambiental en México." In *Daños a la salud por contaminación atmosférica*, edited by Favio Gerardo Rico Méndez, Rafael López Castañares, Ezequiel Jaimes Figueroa, 420–33. Toluca: Universidad Autónoma del Estado de México, 2001.

Palacios, Juan. "Memoria sobre incendio del pozo de petroleo de 'Dos Bocas.'" *Boletín de la sociedad Mexicana de Geografia y Estadistica, quinta época* 3 (1908): 3–40.

Palmer, Steven. *From Popular Medicine to Medical Populism: Doctors, Healers, and Public Power in Costa Rica, 1800–1940*. Durham, NC: Duke University Press, 2003.

Parra, Guillermo. "Algunas consideraciones sobre el hipnotismo desde el punto de vista terapéutico." *La Escuela de Medicina* 13, no. 21 (April 1, 1896): 463–66.

Parra, Porfirio. "The General Character of the Positive Method." In *Nuevo sistema de lógica inductiva y deductiva*. Mexico City: Tipografía Ecónomica, 1903.

Peon de Valle, Juan. "Sesión ordinaria del día 5 de Julio de 1901." *El Observador Médico* 1, no. 11 (September 1, 1901): 173–76.

Pérez Bertruy, Ramona. "La constitución de paseos y jardines públicos modernos en la Ciudad de México durante el Porfiriato: Una experiencia social." In *Los espacios públicos de la ciudad, siglos XVII y XIX*, edited by Carlos Aguirre Anaya, Marcela Cávalas, María Amparo Pos, 314–34. Mexico City: Casa Juan Pablo, 2002.

Perló Cohen, Manuel. *El paradigma porfiriano: historia del desagüe del valle de México*. Mexico City: Universidad Autónoma de México, 1999.

Piccato, Pablo. *City of Suspects: Crime in Mexico City, 1900–1931*. Durham, NC: Duke University Press, 2001.

Pilcher, Jeffrey. "Mad Cowmen, Foreign Investors and the Mexican Revolution." *Journal of Iberian and Latin American Studies* 4, no. 1 (1998): 1–15.

Pilcher, Jeffrey M. *The Sausage Rebellion: Public Health, Private Enterprise, and Meat in Mexico City, 1890–1917*. Albuquerque: University of New Mexico Press, 2006.

Poniatowska, Elena. *Nothing, Nobody: The Voices of the Mexico City Earthquake*. Philadelphia, PA: Temple University Press, 1995.

Porter, Roy. "The Patient's View: Doing Medical History from Below." *Theory and Society* 14, no. 2 (1985): 175–98.

Posalagua, M. A. *Estudio para la formación de hospitales generales en la Ciudad de México*. Mexico City: Imprenta de Comercio de Nabor Chávez, 1874.

Pyne, Stephen J. *World Fire: The Culture of Fire on Earth*. Seattle: University of Washington Press, 1997.

Rangel M., Carlos. *Historia del cuerpo de bomberos de Panamá*. Panamá: Imprenta Nacional, 1962.

Recopilación de leyes, decretos y providencias de los poderes legislativo y ejecutivo de la unión. Vol. 18. Mexico City: Imprenta del Gobierno en Palacio, 1874.

Reggiani, Andrés. "De rastacuero a expertos: Modernización, diplomacia cultural y circuitos académicos transnacionales, 1870–1940." In *Los lugares del saber: Contextos locales y redes transnacionales en la formación del conocimiento moderno*, edited by Ricardo Salvatore, 159–87. Rosario, Argentina: Beatriz Viterbo, 2007.

Reguillo, Rossana. "The Social Construction of Fear: Urban Narratives and Practices." In *Citizens of Fear: Urban Violence in Latin America*, edited by Susana Rotker, 187–206. New Brunswick, NJ: Rutgers University Press, 2002.

Reverdin, Jacques Louis. "Greffe épidermique." *Bulletin de la Société Impériale de Chirurgie Paris* 10 (1869): 511–15.

Rivera Cambas, Manuel. *México pintoresco, artístico y monumental*. Vol. 1. Mexico City: Editorial Valle de México, 1972.

Roberts, Owen. "The Politics of Health and the Origins of Liverpool's Lake Vyrnwy Water Scheme, 1871–92." *Welsh History Review/Cylchgrawn Hanes Cymru* 20, no. 2 (2000): 308–35.

Rodríguez, Julia. *Civilizing Argentina: Science, Medicine, and the Modern State*. Chapel Hill: University of North Carolina Press, 2006.

Rodríguez, Martha Eugenia. *La escuela nacional de medicina, 1833–1910*. Mexico City: Universidad Nacional Autónoma de México, 2008.

Rodríguez de Romo, Ana Cecilia. "La ciencia pasteuriana a través de la vacuna antirrábica: el caso mexicano." *Dynamis* 16 (1996): 291–316.

Rodriguez-Santana, Ivette. "Conquests of Death: Disease, Health and Hygiene in the Formation of a Social Body (Puerto Rico, 1880–1929)." PhD diss., Yale University, 2005.

Rosenberg, Charles E., and Janet Golden. *Framing Disease: Studies in Cultural History*. New Brunswick, NJ: Rutgers University Press, 1992.

Rosenberg, Nathan. "The Historiography of Technical Progress." In *Inside the Black Box: Technology and Economics*, edited by Nathan Rosenberg, 3–33. Cambridge: Cambridge University Press, 1982.

Rosenberg, Nathan. "The International Transfer of Technology: Implications for the Industrialized Countries." In *Inside the Black Box: Technology and Economics*, edited by Nathan Rosenberg, 245–79. Cambridge: Cambridge University Press, 1982.

Rosenthal, Anton Benjamin. "Spectacle, Fear, and Protest: A Guide to the History of Urban Public Space in Latin America." *Social Science History* 24, no. 1 (2000): 33–73.

Rosner, David, and Gerald Markowitz. *Deceit and Denial: The Deadly Politics of Industrial Pollution*. Berkeley: University of California Press, 2002.

Ross, Paul. "Mexico's Superior Health Council and the American Public Health Association: The Transnational Archive of Porfirian Public Health, 1887–1910." *Hispanic American Historical Review* 89, no. 4 (2009): 573–602.

Rothschild, Henri de. *Tratamiento de las Quemaduras por el método céreo (cura por la ambrina)*. Translated by Dr. D. José de Sard. Barcelona: P. Salvat, 1919.

Royal Insurance Company. *Index to Classification of Fire Hazards: North America*. S.L.: S.N., 1897. Hagley Library.

Sá Almeida, Anna Beatriz de. "A Associação Brasileira de Medicina do Trabalho: Locus do processo de constituição da especialidade medicina do trabalho no Brasil na década de 1940." *Ciência e Saúde Coletiva* 13, no. 3 (2008): 869–77.

Sackman, Douglas Cazaux. *Orange Empire: California and the Fruits of Eden*. Berkeley: University of California Press, 2005.

Salazar, Luís. "On the Distribution of Water in the City of Mexico." *International Engineering Congress* 30 (1893): 344.

Sánchez Flores, Ramón. *Historia de la tecnología y la invención: introducción a su estudio y documentos para los anales de la técnica*. Mexico City: Fomento Cultural Banamex, 1980.

Santiago, Myrna I. *The Ecology of Oil: Environment, Labor, and the Mexican Revolution, 1900–1938*. Studies in Environment and History. Cambridge: Cambridge University Press, 2006.

Santiago, Myrna. "Rejecting Progress in Paradise: Huastecs, the Environment, and the Oil Industry in Veracruz, Mexico, 1900–1935." *Environmental History* 3, no. 2 (1998): 169–88.

Schmookler, Jacob. *Invention and Economic Growth.* Cambridge, MA: Harvard University Press, 1966.

Schwarcz, Lilia Moritz. *The Spectacle of the Races: Scientists, Institutions, and the Race Question in Brazil.* Translated by Leland Guyer. New York: Hill and Wang, 1999.

Sedrez, Lise. "Latin American Environmental History: A Shifting Old/New Field." In *The Environment and World History,* edited by Edmund Burke III and Kenneth Pomeranz, 255–75. Berkeley: University of California Press, 2009.

Segura y Tornel, Adrian. "Breves consideraciones acerca del tratamiento de las úlceras cutáneas." MD thesis, Escuela de Medicina, 1874.

Sellers, Christopher C. *Hazards of the Job: From Industrial Disease to Environmental Health.* Chapel Hill: University of North Carolina Press, 1997.

Sicherman, Barbara. *Alice Hamilton: A Life in Letters.* Cambridge, MA: Harvard University Press, 1984.

Sinclair, Bruce. *A Centennial History of the American Society of Mechanical Engineers, 1880–1980.* Toronto: University of Toronto Press, 1980.

Sinclair, Upton. *The Jungle.* New York: Doubleday, Jabber & Company, 1906.

Slaton, Amy. *Reinforced Concrete and the Modernization of American Building, 1900–1930.* Baltimore, MD: Johns Hopkins University Press, 2001.

Smith, Bruce L. R. *The Advisers: Scientists in the Policy Process.* Washington, DC: Brookings Institution, 1992.

Smith, Merritt Roe, and Leo Marx. *Does Technology Drive History? The Dilemma of Technological Determinism.* Cambridge, MA: MIT Press, 1994.

Smyth, Frederick. *Discourses and Letters Commemorative of Emily Lane, Wife of Ex-Gov. Frederick Smyth.* Manchester, NH: John B. Clarke, 1885.

Sosa, Secundino E. "Su higiene, sus enfermedades." MD thesis, La Escuela Nacional de Medicina, 1888.

Soto Laveaga, Gabriela, and Claudia Agostoni. "Science and Public Health in the Century of Revolution." In *A Companion to Mexican History and Culture,* edited by William H. Beezley, 561–74. Blackwell Companions to World History. Malden, MA: Wiley-Blackwell, 2011.

Stephan, Nancy Leys. *The Hour of Eugenics: Race, Gender, and Nation in Latin America.* Ithaca, NY: Cornell University Press, 1991.

Steward, E. Burton. *A Fire Department Training School.* New York: New York Underwriters Agency, 1896.

Stonich, Susan. "International Tourism and Disaster Capitalism: The Case of Hurricane Mitch in Honduras." In *Capitalizing on Catastrophe: Neoliberal*

Strategies in Disaster Reconstruction, edited by Nandini Gunewardena and Mark Schuller, 47–68. Toronto: AltaMira Press, 2008.

Supple, Barry. "Insurance in British History." In *The Historian and the Business of Insurance*, edited by Oliver M. Westall, 1–8. Manchester: Manchester University Press, 1984.

Suzigan, Wilson. *Industria brasileira: Origem e desenvolvimento.* São Paulo: Editoria Brasiliense, 1986.

Sze, Julie. *Noxious New York: The Racial Politics of Urban Health and Environmental Justice.* Cambridge, MA: MIT Press, 2007.

Talbot, Winthrop. "Some Economic Aspects of Factory Hygiene." *American Journal of Public Health* 2, no. 10 (1912): 773–75.

Tamayo, Jorge L. *Breve reseña sobre la escuela nacional de ingeniera.* Mexico City: Imprenta La Esfera, 1958.

Tenorio-Trillo, Mauricio. *Mexico at the World's Fairs: Crafting a Modern Nation.* Berkeley: University of California Press, 1996.

Terrés, José. "Lecciones del Dr. José Terrés. Primera Lección." In *Anales de la Escuela Nacional de Medicina. Parte Médica. Año I. 1904–1905*, 14–42. Mexico City: Tipografía de los sucs. de Francisco Díaz de Leon, 1905.

Terry, Thomas Philip. *Terry's Mexico: Handbook for Travellers.* London: Gay and Hancock, 1911.

Tomes, Nancy. "The Private Side of Public Health: Sanitary Science, Domestic Hygiene and the Germ Theory, 1870–1900." *Bulletin of the History of Medicine*, no. 64 (1990): 509–39.

Torre, Juan de la. *Legislación de patentes y marcas.* Mexico City, 1903.

Trabulse, Elías. *Las patentes de invención durante el siglo XIX en México.* 3rd series. No. 34. Mexico City: Boletín del Archivo General de la Nación, 1988.

Ueyama, Takahiro. *Health in the Marketplace: Professionalism, Therapeutic Desires, and Medical Commodification in Late-Victorian London.* Palo Alto, CA: Society for the Promotion of Science and Scholarship, 2010.

Ulloa, Miguel. *Memoria de la primera exposición en la capital del estado de México, Toluca.* México: Tipografía Literaria del Filomeno Mata, 1883.

Usher, Abbott Payson. *A History of Mechanical Inventions.* 2nd ed. Cambridge, MA: Harvard University Press, 1954.

Valenzuela, J. "De la asistencia médica a los enfermos pobres a domicilio." *La Escuela de Medicina* 3, no. 24 (June 15, 1882): 329–31.

Vanderwood, Paul J. *Disorder and Progress: Bandits, Police, and Mexican Development.* Wilmington, DE: Scholarly Resources, 1992.

Vega, Marta de la. *Evolucionismo versus positivismo: estudio teórico sobre el positivismo y su significación en América Latina.* Caracas, Venezuela: Monte Avila Editores Latinoamericana, 1998.

Vergara, Angela. "The Recognition of Silicosis: Labor Unions and Physicians in the Chilean Copper Industry, 1930s–1960s." *Bulletin of the History of Medicine* 79, no. 4 (2005): 723–48.

Victoria, José Guadalupe. "Noticias sobre la antigua plaza y el mercado del volador de la Ciudad de México." *Anales del Instituto de Investigaciones Estéticas* 16, no. 62 (1991): 69–91.

Vidal y Flor, Luís A. *Colección de Leyes Federales Vigentes sobre Instituciones de Crédito, Ferrocarriles, Compañías de Seguros, Almacenes generals de Depósito y varias Circulares importantes recientes.* Mexico City: Herrero Hermanos, 1900.

Vigarello, Georges. *Lo limpio y lo sucio: La higiene del cuerpo desde la edad media.* Madrid: Alianza Editorial, 1985.

Villarello, Juan D. "El pozo de petróleo de 'Dos Bocas.'" *Boletín del Instituto Geológico de México* 3 (1909): 17–21.

Vitz, Matthew. "Revolutionary Environments: The Politics of Nature and Space in the Valley of Mexico, 1890s–1940s." PhD diss., New York University, 2010.

Vitz, Matthew. "'To Save the Forests': Power, Narrative, and Environment in Mexico City's Cooking Fuel Transition." *Mexican Studies/Estudios Mexicanos* 31, no. 1 (2015): 125–155.

Wakild, Emily. "Naturalizing Modernity: Urban Parks, Public Gardens and Drainage Projects in Porfirian Mexico City." *Mexican Studies/Estudios Mexicanos* 23, no. 1 (2007): 101–23.

Watson, Irving A. "The Republic of Mexico—Medicine Curative and Preventive." *The Sanitarian, a Monthly Magazine Devoted to the Preservation of Health, Mental and Physical Culture* 29, no. 272 (1892): 116–26.

Wecker, Louis de. "De la greffe dermique en chirurgie oculair." *Annales d'oculistique* 68 (1872): 62–71.

Wermiel, Sara E. *The Fireproof Building: Technology and Public Safety in the Nineteenth-Century American City.* Studies in Industry and Society. Baltimore, MD: Johns Hopkins University Press, 2000.

Zoldivar, Luis G. *Apendice a la recopilación de leyes del año de 1859 formado por Luis G. Zoldivar.* Mexico City: Imprenta de A. Boix, 1865.

INDEX